New York Yacht Club

Code of Yachting Signals

New York Yacht Club

Code of Yachting Signals

ISBN/EAN: 9783337404024

Printed in Europe, USA, Canada, Australia, Japan

Cover: Foto ©berggeist007 / pixelio.de

More available books at **www.hansebooks.com**

CODE OF Yachting Signals

Compiled under
THE DIRECTION OF
The New York Yacht Club.
1874.

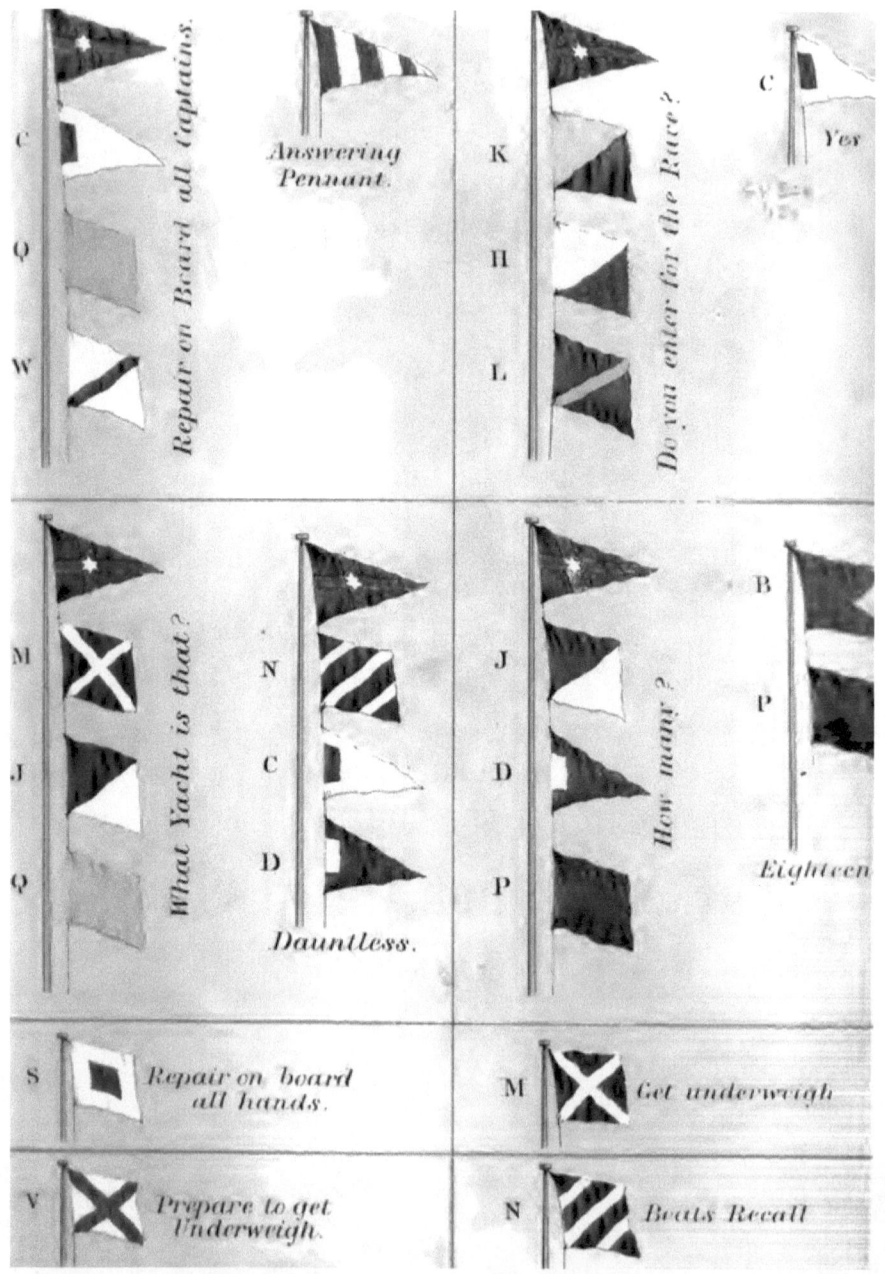

N. Y. Yacht Club  Signal.

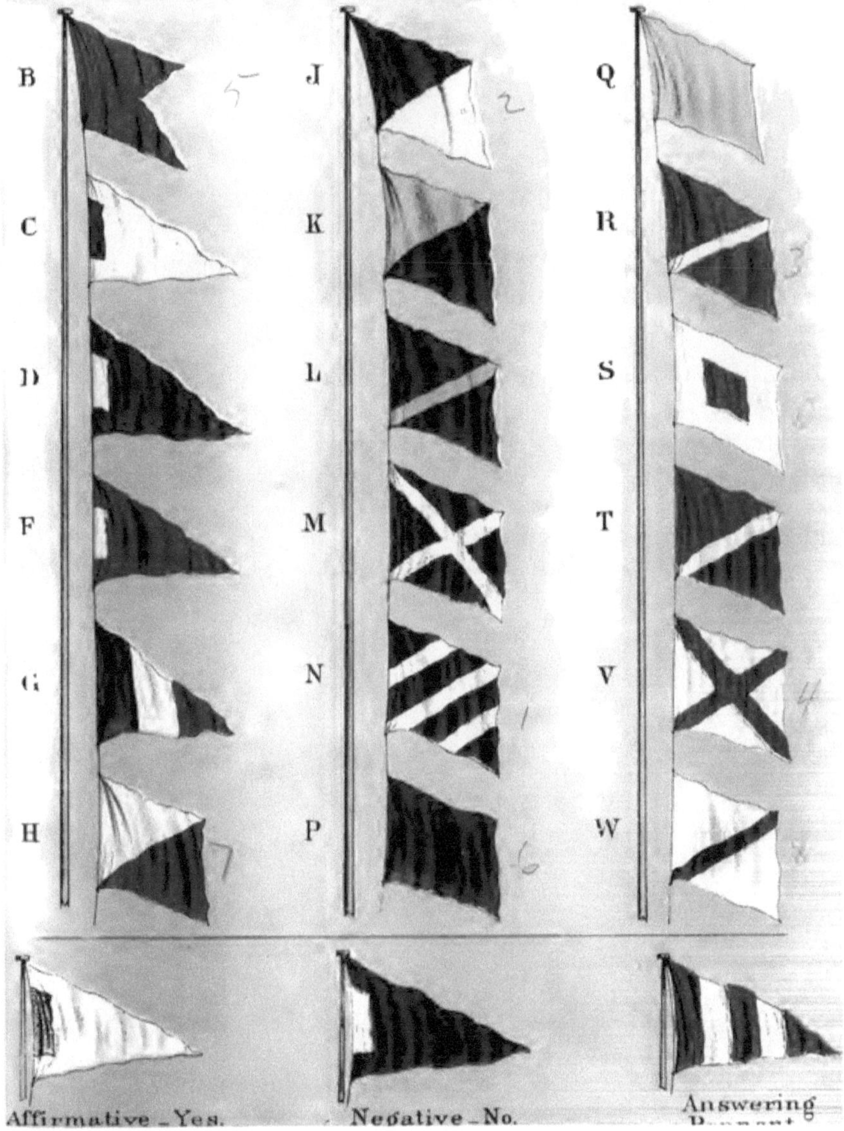

B	J	Q
C	K	R
D	L	S
F	M	T
G	N	V
H	P	W

Affirmative – Yes.   Negative – No.   Answering Pennant.

NIGHT SIGNAL CHART.

(COSTON'S TELEGRAPHIC SYSTEM.)

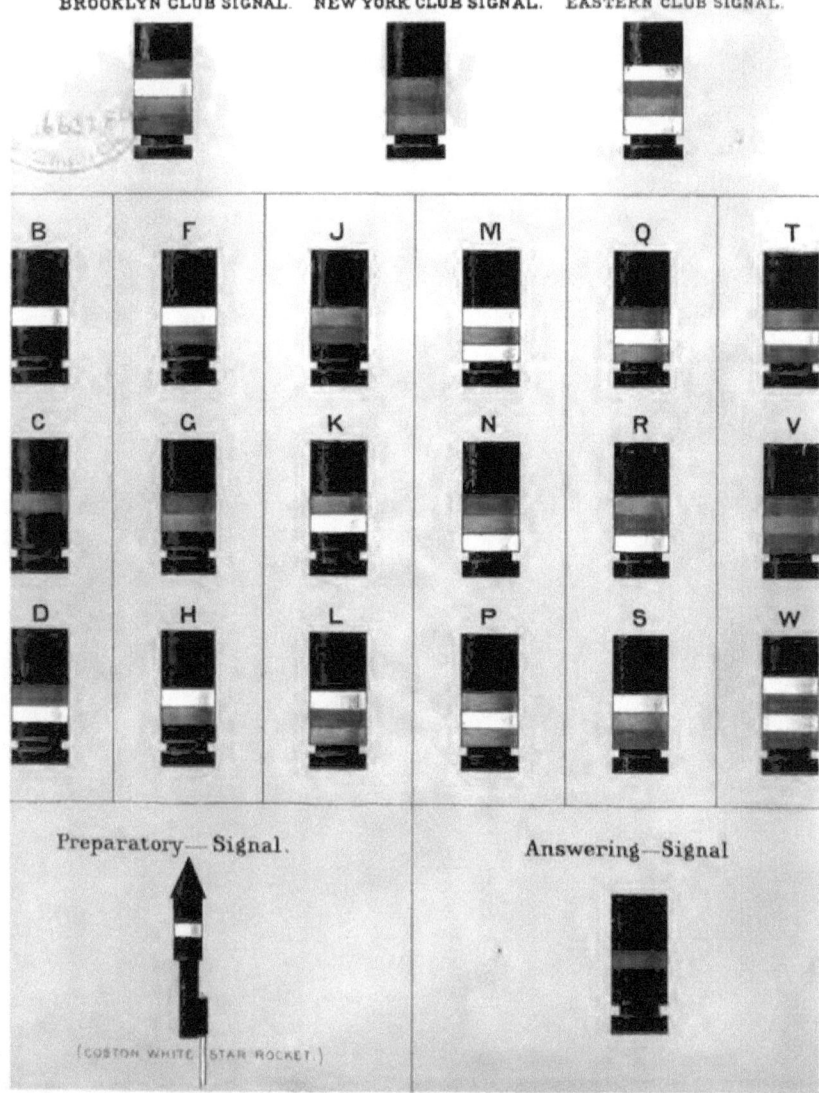

INDEX.

PREPARATORY. .... 3 | SQUADRON EVOLUTIONS 4

CONVERSATIONAL CODE.

A

	PAGE
ABIDE	6
ABLE	6
ABOARD	6
ABOUND	7
ABSENCE	7
ABSENT	7
ACCEPT	8
ACCEPTED	8
ACCIDENT	8
AFLOAT	9
ANCHOR	9
ANCHORAGE	10
ANSWER	11
APPOINT	11
APPOINTED	12
APPEAL	12
APPEALED	13
ARRIVE—D	13
ASHORE	13

	PAGE
ASLEEP	14
ASSIST	14
ASSISTANCE	15
ATTEND	15
ATTENTION	16
ATTENTIVE	16
AVAIL	16
AWAKE	17
AWARD	17
AWARDED	17

B

	PAGE
BACK	18
BAD	18
BADLY	19
BALL	19
BAND	20
BAROMETER	20
BAY	20

INDEX.

	PAGE		PAGE
BED	21	DANCE	38
BEFORE	21	DANCING	38
BEG	21	DANGER	39
BELIEVE	22	DANGEROUS	39
BELL	22	DARK	39
BEST	22	DATE	40
BETTER	23	DAY	42
BEWARE	24	DELAY—ED	43
BOAT	24	DINE—D	43
BOATS	24	DINING	44
BOUND	25	DINNER	44
BREAKFAST	25	DRESS—ED	45
BUOY	26	DRY	46
		DURING	46

C

E

	PAGE		PAGE
CABLE	26	EACH	47
CAPTAIN	26	EAST	47
CAPTAINS	27	EASTWARD	47
CHALLENGE	27	ENDANGER	48
COME	28	ENDEAVOR—ED	48
COMING	28	ENFORCE—D	48
COMMODORE	28	ENGAGED	49
COMMUNICATE	30	ENGAGEMENT	49
COMMUNICATION	30	ENOUGH	49
COMPANY	30	ENSIGN	50
COMPASS	31	ENTER	50
COURSE	31	ENTERTAIN	51
CURRENT	37	ENTERTAINED	51
		ENTERTAINMENT	51

D

		ENTRANCE	51
DAMAGE	37	ERROR	52
DAMAGES	37	ESCAPE—D	52

INDEX.

	PAGE
EVENING	53
EXCELL—ED	53
EXCELLENT	53
EXECUTE—D	54

F

FAST	54
FATHOMS	55
FAVOR	55
FEAR	55
FEARFUL	56
FEEL	56
FEELING	56
FEET	57
FINE	57
FIREWORKS	57
FISH	57
FISHING	57
FISHING-LINES	58
FISHERMAN	58
FOG	58
FRIGHTENED	59
FROM	59
FUTURE	59

G

GAFF	60
GAIN	60
GLAD	60
GO—ING	61
GUESTS	61
GUN	61

H

	PAGE
HALLIARDS	62
HAPPEN	62
HARBOR	63
HEAD	63
HEAR	64
HOUR	64
HULL	65
HURRY	65

I

IF	66
IMAGINE—D—ING	66
IMPERIL	67
IMPORTANCE	67
IMPORTANT	67
IMPROPER	67
IMPRUDENT	68
INABILITY	68
INFORMED	68
INFORMATION	69
INJURE	70
INJURY	70
INJURE—D	71
INSTANT—LY	71
INSTRUCTIONS	71
INVITATION	72
INVITE	73
IT	73

J

JOIN	73
JOKE—ING	74
JUDGES	75

INDEX.

K

	PAGE
Kedge	75
Keep	76
Kindness	76
Know—n	76

L

Land	77
Last	78
Late	79
Lead	79
Leak	80
Least	80
Leave—ing	80
Lend	81
Liable	82
Liability	82
Like—ly	82
Line	83
Little	83
Live	84
Loss	84
Lost	84
Luck—y	85
Lunch	86
Lurch	86

M

Made	86
Make	87
Many	87
Mast	88
Matter	89
Medicine	89

	PAGE
Men	89
Message	90
Met	90
Minutes	90
Miss—ed	91
Mistake	91
Month	91
Much	92

N

Name	93
Nations	93
Near	94
New	94
News	94
Night	95
North—erly	95
Numerals	96

O

Of	96
Off	96
Often	97
Opinion	98
Opportunity	98
Orders	99
Other	99
Overboard	100

P

Papers	100
Please	101

INDEX.

	PAGE
Pleasure	101
Postpone—d	102
Prepare—d	102
Preparation	102
Preparing	103
Prevent—ed	103

Q

Question	103
Quickly	104

R

Race	104
Rain—ing—y	106
Rather	107
Reason—s	107
Reflect—ed—ion	108
Rest—ed—ing	108
Return—ed—ing	109
Right—ed	109
Rocks	110
Rudder	110
Run—ning	111

S

Safe—ly	111
Sail	112
Sails	113
Salute	114
Same	114
Save—d	115
Say	115
Sea	115

	PAGE
See—n	116
Shall	116
Sick—ness	116
Sight	117
Signal—s—ed	117
Since	118
Slip—ped	119
So	119
Soundings	119
South—erly—ward	120
Steam—er—s	120
Supper	121
Supply—ies	121
Surgeon	122

T

Tack	122
Talk	122
Tell	123
Think—ing	123
Tide	124
Time	124

U

Unable	125
Underweigh	125
Unwilling	126
Use—ful	126

W

Want	127
Was	127
Water	127

INDEX.

	PAGE		PAGE
Way	128	Wind	132
Wear	128	Wrong	132
Weather	128		
West—erly—ward	130	**Y**	
Where	131	Yacht	132
Why	131	Year	133
Will—ing	131	You—r—s	133

YACHT NUMBERS.

	PAGE		PAGE
Schooners	136	Steamers	142
Sloops	140	Names of Places	144

GENERAL DIRECTIONS

FOR THE USE OF THE

YACHTING CODE OF SIGNALS.

I.—THE DAY SIGNALS.

THE FLAGS used are those of the *Roger's Commercial Code*. When using the Yachting Code, the Club Flag will serve as the distinguishing pennant, and in all cases, except in the "preparatory signals" and "numerals," it must be hoisted over the signal flags. Answers will, as far as practicable, be made by the flags representing Yes and No. Except in the case hereinafter noted, every yacht in the squadron will run up her answering pennant as soon as the signal on the Flag-ship is understood, and will keep it flying until the signals are taken down.

When the commanding officer desires to signal a particular yacht, the number of the latter will be set first, and the remainder of the squadron will be excused from attention to signals until another gun is fired on the Flag-ship.

II.—THE NIGHT SIGNALS.

The signals used are *Coston's Yachting Signals*. The night Code is the same as the day code. Signalling at night will be done as follows:

GENERAL DIRECTIONS.

First—A rocket will be sent up from the Flag-ship (or yacht desiring to signal another) to attract the attention of the squadron (or other yacht).

Second—The signal required will then be made by burning in rapid succession the lights representing the letters composing the signal.

Third—The yacht or yachts signalled will then reply with the "answering signal" light to indicate that the signal is understood, and if a further answer is necessary it will be made by burning the lights representing the letters indicating the answer.

Fourth—A yacht meeting another yacht at night can ascertain her name, and that of her club, by first burning the light representing the "Club Signal," and then the lights indicating her number, to which the yacht signalled will reply in the same manner; but in all cases a rocket should first be sent up to attract attention.

Care should be taken by the person reading a night signal, not to look at the light burning on his own yacht, for otherwise he will find it difficult to distinguish the colors on the signalling yacht.

NOTE.

The "preparatory signals" and "numerals" will be made *without* the club flag over them. The yacht numbers commence with N. Names of places commence with P. All other signals with the club flag over them are in the "Squadron Evolutions" and "Conversational Code."

Squadron Evolutions.

First order of Sailing.

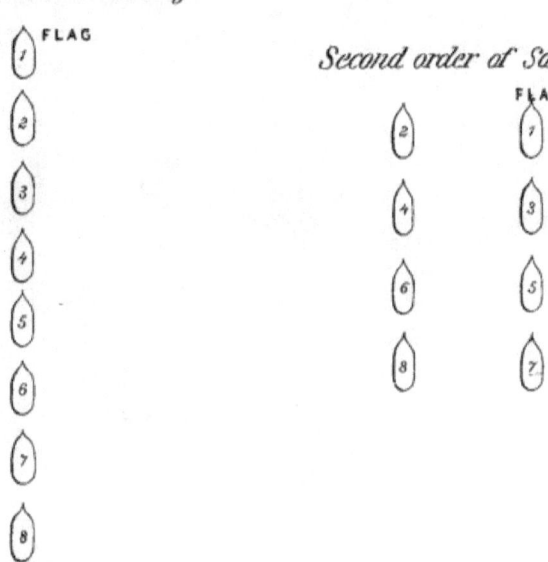

Second order of Sailing.

Third order of Sailing.

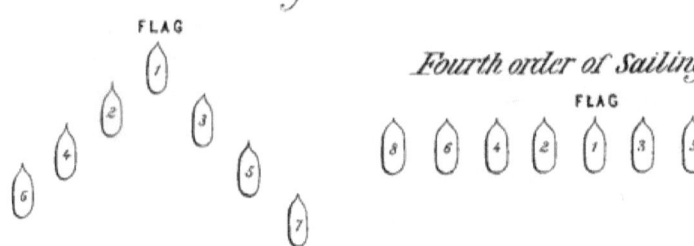

Fourth order of Sailing.

Squadron Evolutions.

Fifth order of Sailing.

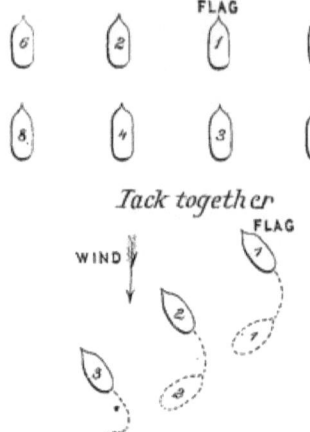

Tack together

Tack in succession in wake of Leading vessel.

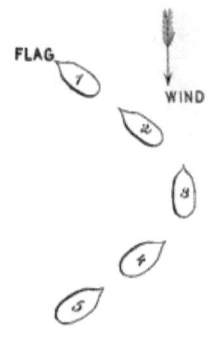

Bear away together at right angles to present course.

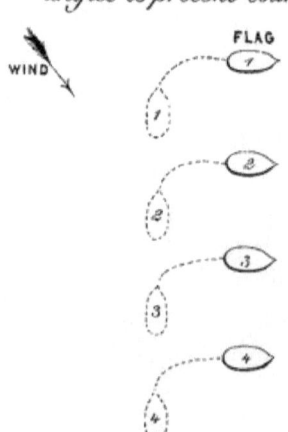

Haul on wind together at right angles to present course.

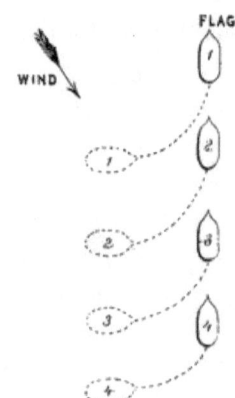

Squadron Evolutions.

From first to second order.

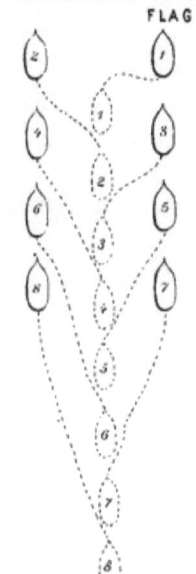

From third to fourth order.

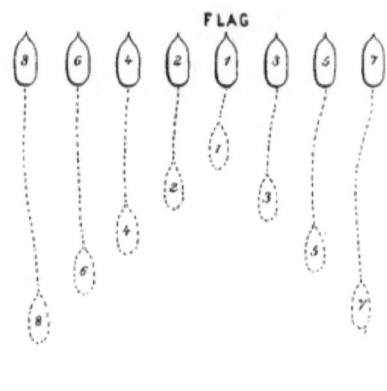

From second to third order.

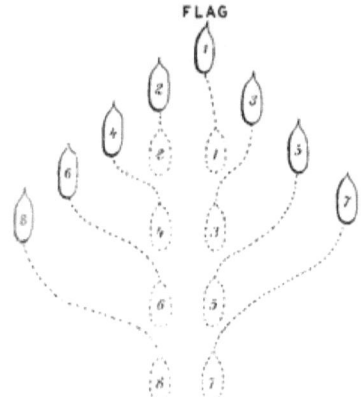

From fourth to fifth order.

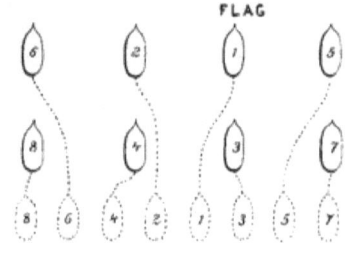

Yachting Code of Signals.

PREPARATORY.

Note.—In making the following signals, the flag alone will be hoisted (with no pennant above it).

S   Repair on board all hands.
V   Prepare to get under weigh.
M   Get under weigh.
N   Boats recall.
C   Yes.
D   No.

Yachting Code of Signals.

SQUADRON EVOLUTIONS.

Yachts will commence to execute an evolution when the signal ordering it is hauled down.

The answering pennant must be hoisted as soon as the signal is seen and understood.

B	C	D	Anchor without regard to order of sailing.
B	C	F	Bear away together at right angle to present course.
B	C	G	Bear up together at right angle to present course.
B	C	H	Bear up in succession in wake of leading vessel.
B	C	J	Bear away in succession in wake of leading vessel.
B	C	K	Close more the order of sailing.
B	C	L	Commodore will lead, other vessels to follow without regard to any particular order.
B	C	M	Disregard my motions.
B	C	N	Disregard all particular order of sailing.
B	C	P	Fill away.
B	C	Q	Follow my motions.
B	C	R	Form the first order of sailing.
B	C	S	Form the second order of sailing.
B	C	T	Form the third order of sailing.
B	C	V	Form the fourth order of sailing.
B	C	W	Form the fifth order of sailing.
B	D	C	Haul on the wind together at right angle to present course.
B	D	F	Haul by the wind on the starboard tack.
B	D	G	Haul by the wind on the port tack.
B	D	H	Heave to.

Yachting Code of Signals.

SQUADRON EVOLUTIONS.—Continued.

B	D	J	Increase distance between port and starboard divisions.
B	D	K	Lessen distance between port and starboard divisions.
B	D	L	Make more sail.
B	D	M	Open more the order of sailing.
B	D	N	Pay more attention to signals.
B	D	P	Shall we shorten sail?
B	D	Q	Shorten sail.
B	D	R	Tack together.
B	D	S	Tack in succession, in wake of leading vessel.
B	D	T	Wear together.
B	D	V	Wear in succession, in wake of leading vessel.

CONVERSATIONAL CODE.

ABIDE.

B F C   You must abide by the decision of the senior officer.

ABLE.

B F D   Are you not able?
B F G   I am able.
B F H   I will, if able.
B F J   I am not able.

ABOARD.

B F K   Have you any visitors aboard?
B F L   What visitors have you aboard?
B F M   Will your visitors come aboard?
B F N   Do you expect visitors aboard?
B F P   Shall I bring them aboard?
B F Q   Will you come aboard?
B F R   Do you wish me to come aboard?
B F S   When would you like to see me aboard?
B F T   As soon as you can come aboard.
B F V   It will be impossible to go aboard.
B F W   Bring them aboard with you.

Yachting Code of Signals.

ABOUND.

B	G	C	The waters abound with fish.
B	G	D	The locality abounds with game.

ABSENCE.

B	G	F	What is the reason of his absence?
B	G	H	Leave of absence is requested.
B	G	J	Leave of absence is granted.
B	G	K	Leave of absence cannot be granted.
B	G	L	You may prolong your absence.
B	G	M	Your absence must be accounted for.
B	G	N	In the absence of the Commodore.
B	G	P	In the absence of the Vice Commodore.
B	G	Q	In the absence of the Rear Commodore.
B	G	R	In the absence of the Captain.
B	G	S	I, in his absence, cannot leave.
B	G	T	His absence is inexcusable.
B	G	V	In the absence of our visitors.
B	G	W	There are satisfactory reasons for his absence.

ABSENT

B	H	C	How long will he be absent?
B	H	D	Our guests are absent.
B	H	F	Our guests will be absent for some time.
B	H	G	He is absent on leave.

ACCEPT.

B	H	J	Will you accept my apology?
B	H	K	Will you accept the invitation?
B	H	L	Will your friends accept the invitation?
B	H	M	Shall be happy to accept your invitation.
B	H	N	It will be impossible to accept your invitation.
B	H	P	Will accept your invitation if possible.
B	H	Q	Cannot accept your invitation under any circumstances.
B	H	R	I will accept with pleasure.
B	H	S	I will accept your excuse.
B	H	T	I cannot accept your excuse.
B	H	V	I accept your apology.

ACCEPTED.

B	H	W	The invitation has been accepted.
B	J	C	The invitation cannot be accepted.
B	J	D	The invitation will be accepted.
B	J	F	Will the invitation be accepted?
B	J	G	The apology is accepted.
B	J	H	The apology cannot be accepted.
B	J	K	I have accepted.

ACCIDENT.

B	J	L	We have had an accident on board.
B	J	M	We trust the accident is not serious.

Yachting Code of Signals.

ACCIDENT.—Continued.

B	J	N	The accident is not serious.
B	J	P	The accident is very serious.
B	J	Q	How did the accident occur?
B	J	R	The accident was the result of carelessness.
B	J	S	The accident was unavoidable.

AFLOAT.

B	J	T	Will you be afloat at low water?
B	J	V	Will you be afloat at high water?
B	J	W	Is your boat afloat?
B	K	C	When will you be afloat?
B	K	D	We shall be afloat at low water.
B	K	F	We shall be afloat at high water.
B	K	G	Our boat is not yet afloat.

ANCHOR.—(See also "Dark.")

B	K	H	May I anchor?
B	K	J	Have you a spare anchor?
B	K	L	Where are we to anchor?
B	K	M	How are we to anchor?
B	K	N	I wish to anchor before dark.
B	K	P	My anchor is foul.
B	K	Q	I have lost my anchor.
B	K	R	Anchor.

ANCHOR.—Continued.

B	K	S	Anchor to windward.
B	K	T	Anchor to leeward.
B	K	V	Anchor ahead.
B	K	W	Anchor astern.
B	L	C	Anchor on my lee quarter.
B	L	D	Anchor on my lee bow.
B	L	F	Anchor on my weather quarter.
B	L	G	Anchor on my weather bow.
B	L	H	Anchor in line abreast and to windward.
B	L	J	Anchor in line abreast and to leeward.

ANCHORAGE.

B	L	K	Reconnoitre and report upon the anchorage.
B	L	M	We wish to shift our anchorage, bottom is foul.
B	L	N	Prepare to shift anchorage.
B	L	P	The anchorage is bad, do not come to.
B	L	Q	The anchorage is very good, and well sheltered.
B	L	R	The anchorage is open, but holding ground is good.
B	L	S	Stand in to the anchorage.
B	L	T	Can you lead to the anchorage?
B	L	V	Is the anchorage good?
B	L	W	Is the anchorage bad?
B	M	C	Can the anchorage be depended upon.

Yachting Code of Signals.

ANSWER.

B	M	D	What answer have you?
B	M	F	Is the answer in the affirmative?
B	M	G	Is the answer in the negative?
B	M	H	Does it require an answer?
B	M	J	Is the answer verbal, or in writing?
B	M	K	The answer is satisfactory.
B	M	L	The answer is not satisfactory.
B	M	N	An immediate answer is requested.
B	M	P	Signal the answer as soon as received.
B	M	Q	That will answer.
B	M	R	It will not answer.
B	M	S	Will answer as soon as possible.
B	M	T	No answer is required.

APPOINT.

B	M	V	What time will you appoint?
B	M	W	Appoint the time to suit yourself.
B	N	C	Appoint a committee.
B	N	D	Appoint a referee.
B	N	F	Appoint a day.
B	N	G	Appoint an hour.
B	N	H	I will appoint a committee.
B	N	J	I will appoint the time.
B	N	K	Appoint a meeting.
B	N	L	Appoint a conference.

APPOINTED.

B	N	M	What time is appointed for the meeting?
B	N	P	The meeting is appointed.
B	N	Q	Will be there at the time appointed.
B	N	R	At the place appointed.
B	N	S	On the day appointed.
B	N	T	At the hour appointed.
B	N	V	The committee will be appointed.
B	N	W	A committee has been appointed.
B	P	C	A referee has been appointed.
B	P	D	A time has been appointed.
B	P	F	The time will be appointed.
B	P	G	The time has not yet been appointed.

APPEAL.

B	P	H	Can I appeal from the decision?
B	P	J	Do you intend to appeal?
B	P	K	There is no appeal, the decision is final.
B	P	L	I intend to appeal from the decision.
B	P	M	I appeal to you.
B	P	N	I must appeal.
B	P	Q	The appeal has been made.
B	P	R	No appeal will be made.
B	P	S	The appeal is not necessary.
B	P	T	'Twould be folly to appeal.
B	P	V	The appeal will be favorably considered.

Yachting Code of Signals.

APPEALED.

B	P	W	Why have they appealed?
B	Q	C	He has appealed.
B	Q	D	He has not appealed.
B	Q	F	He has appealed for good reasons.
B	Q	G	The decision should have been appealed from.

ARRIVE-D.

B	Q	H	When will they arrive?
B	Q	J	They will arrive.
B	Q	K	He may arrive soon.
B	Q	L	He will not arrive until late.
B	Q	M	The mail has arrived.
B	Q	N	The mail has not arrived.
B	Q	P	When the mail has arrived, let me know.
B	Q	R	Our guests have arrived.
B	Q	S	The yacht has arrived at ———.

ASHORE.

B	Q	T	Are you going ashore?
B	Q	V	When are you going ashore?
B	Q	W	Have the party gone ashore?
B	R	C	Is there trouble in getting ashore?
B	R	D	Will you join us ashore?
B	R	F	I am going ashore.

Yachting Code of Signals.

ASHORE—Continued.

B	R	G	I am going ashore at ———.
B	R	H	There is a yacht ashore at ———.
B	R	J	You can go ashore without difficulty.
B	R	K	Do not attempt to go ashore, landing is dangerous.
B	R	L	I will join you ashore.

ASLEEP.

B	R	M	Are you all asleep?
B	R	N	They are all asleep.
B	R	P	Our guests are asleep.
B	R	Q	Are your guests asleep?

ASSIST.

B	R	S	Can you assist me?
B	R	T	Do you wish me to assist you?
B	R	V	Will they assist us?
B	R	W	I will assist you if possible.
B	S	C	It will be impossible to assist you.
B	S	D	Assist him if possible.
B	S	F	I am unable to assist.

ASSISTANCE.

B	S	G	Shall I go to her assistance?
B	S	H	Are you in need of assistance?
B	S	J	Did the assistance arrive in time?
B	S	K	Go to her assistance.
B	S	L	I am in need of assistance.
B	S	M	Assistance is required at once.
B	S	N	Assistance was in time.
B	S	P	Assistance arrived too late.
B	S	Q	Thanks for your assistance.
B	S	R	Assistance is impossible.
B	S	T	Assistance should be sent immediately.
B	S	V	They are not in need of assistance.
B	S	W	The assistance is (was) fully appreciated.

ATTEND.

B	T	C	Must (shall) I attend?
B	T	D	Are we required to attend?
B	T	F	Are we expected to attend?
B	T	G	Will you attend?
B	T	H	Attend to the order (request.)
B	T	J	You will have to attend.
B	T	K	I hope you will attend.
B	T	L	You need not attend.
B	T	M	You are expected to attend.
B	T	N	You are required to attend.
B	T	P	You are not required to attend.

ATTENTION.

B	T	Q	Strict attention to the order (request) is required.
B	T	R	Show him every possible attention.
B	T	S	I will show him every attention.
B	T	V	Do they expect (require) much attention?
B	T	W	Thanks for your attention.
B	V	C	Your attention is appreciated.
B	V	D	Do not attract attention.
B	V	F	He is attracting too much attention.
B	V	G	Particular attention must be paid to signals.
B	V	H	The attention paid has been noticed.

ATTENTIVE.

B	V	J	Would you have me more attentive?
B	V	K	You are very attentive.
B	V	L	How very attentive you are.
B	V	M	You are not attentive to the orders (requests, signals).
B	V	N	You should be more attentive.
B	V	P	You are an attentive person.
B	V	Q	You are attentive to our pleasure.
B	V	R	Be more attentive.

AVAIL.

B	V	S	Of what avail will it be?
B	V	T	It is of no avail.
B	V	W	It will avail him nothing.
B	W	C	I will avail myself of your offer.

Yachting Code of Signals.

AWAKE.

B	W	D	Are you awake yet?
B	W	F	Will you awake him?
B	W	G	We are all awake.
B	W	H	We are hardly awake yet.
B	W	J	It is time he was awake.
B	W	K	When he is awake inform us.
B	W	L	I was awake all night.
B	W	M	It is too early to awake him.
B	W	N	Be sure to awake him in time.
B	W	P	He will awake in time.
B	W	Q	Awake him as quickly as possible.
B	W	R	We are awake to the danger.

AWARD.

B	W	S	Has the award been made?
B	W	T	When will the award be made?
B	W	V	What was the nature of the award?
C	B	D	The award has not been made.
C	B	F	I do not know when the award will be made.
C	B	G	I do not know the nature of the award.

AWARDED.

C	B	H	To what yacht was the prize awarded?
C	B	J	Has the prize been awarded?

AWARDED.—Continued.

C	B	K	Will there be a prize **awarded**?
C	B	L	Why was it so awarded?
C	B	M	Was any salvage awarded?
C	B	N	The prize has not been awarded.
C	B	P	There will be a prize awarded.
C	B	Q	**It is not known** to what yacht the prize is awarded.
C	B	R	We think it was awarded unfairly.
C	B	S	It was awarded fairly.

B

BACK.

C	B	T	Shall I go back?
B	B	V	Shall we go back?
C	B	W	I am going back.
C	D	B	You must go back and try it again.

BAD.

C	D	F	You have kept a bad lookout.
C	D	G	We have kept a bad lookout.
C	D	H	The lookout was bad.
C	D	J	You are in bad luck.
C	D	K	We are in bad luck.

BADLY.

C	D	L	You appear to work badly.
C	D	M	You appear to be badly worked.
C	D	N	We are badly trimmed.
C	D	P	Are you badly trimmed?
C	D	Q	It was very badly executed.
C	D	R	The manœuvre was badly executed.
C	D	S	Do you think the manœuvre was badly executed?
C	D	T	You have behaved badly.

BALL.

C	D	V	Are you going to the ball to-night?
C	D	W	Where is the ball to be given?
C	F	B	Is there a ball on shore to-night?
C	F	D	Will you accompany us to the ball?
C	F	G	Will the ball be postponed?
C	F	H	Have you an invitation to the ball?
C	F	J	Will there be a ball on board the man-of-war?
C	F	K	There is a ball on shore to-night.
C	F	L	The ball is to be given at ———.
C	F	M	I am unable to attend the ball.
C	F	N	The ball has been postponed.
C	F	P	I have received no invitation to the ball.
C	F	Q	There is to be a ball on board the man-of-war.

BAND.

C	F	R	Have you a band on board?
C	F	S	Is the band engaged to-day?
C	F	T	Is the band engaged this evening?
C	F	V	Will you give us some music from the band?
C	F	W	Would you like to hear the band?
C	G	B	Will the band play this evening?
C	G	D	I have a band on board.
C	G	F	The band is at your service.
C	G	H	The band cannot be spared.
C	G	J	The band is on shore.
C	G	K	You have a fine band.
C	G	L	The band will perform this evening.

BAROMETER.

C	G	M	How is your barometer?
C	G	N	Barometer is low and falling.
C	G	P	Barometer is high and rising.
C	G	Q	Barometer is steady.

BAY.

C	G	R	Shall I stand into the bay?
C	G	S	It is a fine bay.
C	G	T	Stand into the bay.
C	G	V	Stand out of the bay.
C	G	W	Do not venture into the bay.

Yachting Code of Signals.

BED.

C	H	B	The bed of the river is muddy.
C	H	D	The bed of the river is hard sand.
C	H	F	The bed of the river is not reliable for holding on.

BEFORE.

C	H	G	Before morning.
C	H	J	Before evening.
C	H	K	Before noon.
C	H	L	Before midnight.
C	H	M	Before sunrise.
C	H	N	Before sunset.
C	H	P	I will go on shore before you.
C	H	Q	You had better go before me.
C	H	R	Will you go before the time appointed?
C	H	S	I will go before the time appointed.
C	H	T	It is better to go before the time appointed.

BEG.

C	H	V	I beg your acceptance.
C	H	W	I beg that you will accept my apology.
C	J	B	I beg your attention.
C	J	D	I beg that you will accompany me.
C	J	F	I must beg to be excused.
C	J	G	I beg pardon.

BELIEVE.

C	J	H	Do you believe it possible?
C	J	K	Do you believe what he says? (they say).
C	J	L	I believe it fully.
C	J	M	I believe him.
C	J	N	You may believe all that is told you.
C	J	P	It is impossible to believe him.
C	J	Q	We cannot believe the rumor (report.)

BELL.

C	J	R	Is there a fog bell there?
C	J	S	Did you hear the fog bell?
C	J	T	There is no fog bell.
C	J	V	I heard the fog bell distinctly.
C	J	W	The fog bell was not heard.
C	K	B	Keep your ears open for the fog bell if weather is thick.
C	K	D	The fog bell has been removed.

BEST.

C	K	F	Where is the best place?
C	K	G	Which is the best place?
C	K	H	When do you think best?
C	K	J	Which do you think best?
C	K	L	What do you think best?
C	K	M	Do you think it best to accept?
C	K	N	Do you think it best to decline?

Yachting Code of Signa

BEST.—Continued.

C	K	P	It is the best place.
C	K	Q	It is the best opportunity you will have.
C	K	R	It is best to be early.
C	K	S	It is best to be late.
C	K	T	It is best to be in time.
C	K	V	It will be best to go as soon as possible.
C	K	W	The move is the best you can possibly make.
C	L	B	It is the best offer I can make.

BETTER.

C	L	D	Could anything be better?
C	L	F	Nothing could be better.
C	L	G	You had better not.
C	L	H	I think it would be better.
C	L	J	I think you had better.
C	L	K	It would be better for him (them).
C	L	M	I hope you will think better of it.
C	L	N	It is impossible to think better of it.
C	L	P	You could not make a better proposition.
C	L	Q	Better late than never.
C	L	R	I hope you are feeling better.
C	L	S	I am much better, thank you.
C	L	T	He is much better.
C	L	V	I have thought better of it.

BEWARE.

C	L	W	Beware of what or whom?
C	M	B	Beware of the danger to leeward.
C	M	D	Beware of the danger ahead.
C	M	F	Beware of him.
C	M	G	You had better beware.
C	M	H	We will beware of you, you are too cunning.
C	M	J	Beware of going too far.
C	M	K	I will beware of him.

BOAT.

C	M	L	Can you lend me a boat?
C	M	N	Can you send a boat?
C	M	P	Did your boat get ashore safely?
C	M	Q	How was the boat race?
C	M	R	I have sent a boat.
C	M	S	A boat will be sent.
C	M	T	The boat capsized.
C	M	V	The boat escaped without injury.
C	M	W	The boat was swamped.
C	N	B	The boat race was a success.
C	N	D	The boat race was a failure.

BOATS.

C	N	F	Have you sent your boats away?
C	N	G	My boats have all been washed away.

Yachting Code of Signals.

BOATS.—Continued.

C	N	H	My boats are all safe.
C	N	J	The boats are all ashore.
C	N	K	Send all boats immediately.
C	N	L	We have sent all boats.
C	N	M	The boats will be here in a short time.
C	N	P	I will send my boats immediately.
C	N	Q	Recall all boats.
C	N	R	The boats were swamped in getting ashore.
C	N	S	The boats got ashore safely.
C	N	T	The boats were wanted badly.
C	N	V	The boats are wanted badly.

BOUND.

C	N	W	Where are you bound?
C	P	B	We are bound to ———.
C	P	D	We are bound on a cruise.

BREAKFAST.

C	P	F	Will you join us at breakfast?
C	P	G	At what hour do you breakfast?
C	P	H	Will there be company at breakfast with you?
C	P	J	I will join you at breakfast.
C	P	K	I cannot join you at breakfast.
C	P	L	We breakfast at ———.

BREAKFAST.—Continued.

C	P	M	There will be company at breakfast.
C	P	N	Breakfast is ready.

BUOY.

C	P	Q	Shall I make fast to the buoy?
C	P	R	Shall we buoy the anchors?
C	P	S	Make fast to the buoy.
C	P	T	Do not make fast to the buoy.
C	P	V	I am made fast to the buoy.
C	P	W	Buoy your anchors and prepare to slip.

C

CABLE.

C	Q	B	Can you spare a cable?
C	Q	D	I have lost my cable.

CAPTAIN.

C	Q	F	Is the Captain on board?
C	Q	G	Where has the Captain gone?
C	Q	H	When will the Captain return?

Yachting Code of Signals.

CAPTAIN.—Continued.

C	Q	J	Is the Captain ashore?
C	Q	K	The Captain is not on board.
C	Q	L	The Captain is visiting another yacht.
C	Q	M	The Captain has gone ashore.
C	Q	N	The Captain is off on an excursion.
C	Q	P	I do not know when the Captain will return.
C	Q	R	The Captain will return shortly.
C	Q	S	The Captain will not return for some time.
C	Q	T	The Captain is asleep.
C	Q	V	The Captain is absent.

CAPTAINS.

C	Q	W	Repair on board all Captains.
C	R	B	Captains will report on board the flag-ship, on coming to anchor.
C	R	D	Captains will meet on board flag-ship at ———.

CHALLENGE.

C	R	F	Has a challenge been sent?
C	R	G	When will the challenge be sent?
C	R	H	A challenge has been sent.
C	R	J	No challenge has been sent.
C	R	K	A challenge will be sent.
C	R	L	No challenge will be sent.

COME.—(See "See.")

C	R	M	When do you wish me to come?
C	R	N	Will you come with me?
C	R	P	I wish you would come.
C	R	Q	I cannot come.
C	R	S	I will come with pleasure.
C	R	T	Come as soon as convenient.
C	R	V	Come as soon as possible, I cannot wait.

COMING.

C	R	W	Are you coming with me?
C	S	B	When are they coming?
C	S	D	Are you coming for us?
C	S	F	He is not coming
C	S	G	We are coming.
C	S	H	We are not coming.
C	S	J	Do not be so long coming.
C	S	K	They are coming soon.
C	S	L	They are not coming until late.
C	S	M	I am coming for you.
C	S	N	They are coming off now.

COMMODORE.

C	S	P	Is the Commodore aboard?
C	S	Q	Is the Commodore ashore?

Yachting Code of Signals.

COMMODORE.—Continued.

C	S	R	Is the Commodore absent?
C	S	T	When will the Commodore return?
C	S	V	Is the Vice Commodore aboard?
C	S	W	Is the Vice Commodore ashore?
C	T	B	Is the Vice Commodore absent?
C	T	D	When will the Vice Commodore return?
C	T	F	Is the Rear Commodore aboard?
C	T	G	Is the Rear Commodore ashore?
C	T	H	Is the Rear Commodore absent?
C	T	J	When will the Rear Commodore return?
C	T	K	The Commodore is absent.
C	T	L	The Commodore is asleep.
C	T	M	The Commodore will return shortly.
C	T	N	The Commodore will not return for some time.
C	T	P	I do not know when the Commodore will return.
C	T	Q	The Vice Commodore is asleep.
C	T	R	The Vice Commodore is absent.
C	T	S	The Vice Commodore will return shortly.
C	T	V	The Vice Commodore will not return for some time.
C	T	W	I do not know when the Vice Commodore will return.
C	V	B	The Rear Commodore is absent.
C	V	D	The Rear Commodore is asleep.
C	V	F	The Rear Commodore will return shortly.
C	V	G	The Rear Commodore will not return for some time.
C	V	H	I do not know when the Rear Commodore will return.

COMMUNICATE.

C	V	J	Will you communicate with ———?
C	V	K	You will communicate with ———.
C	V	L	I will communicate with ———.
C	V	M	I cannot communicate with ———.

COMMUNICATION.

C	V	N	Have you any communication for me?
C	V	P	I have a communication for you.
C	V	Q	I have no communication for you.
C	V	R	Have you been in communication with the ———?
C	V	S	I have not been in communication with the ———.
C	V	T	I have been in communication with the ———.
C	V	W	Place yourself in communication with the ———.

COMPANY.

C	W	B	Have you been in company with any yacht?
C	W	D	Shall I keep company with you?
C	W	F	Have you company on board?
C	W	G	I have been in company with the ———.
C	W	H	I have not seen a yacht.
C	W	J	We parted company on the ———.
C	W	K	You will keep company with me.
C	W	L	They have company on board the yacht.
C	W	M	We have company on board to-day.

COMPANY.—Continued.

C	W	N	I have no company on board.
C	W	P	My company has left.
C	W	Q	You are in good company.
C	W	R	I am in good company.
C	W	S	You are in suspicious company.
C	W	T	I am in suspicious company.
C	W	V	Shall be very glad to have your company.
D	B	C	Part company.

COMPASS.

D	B	F	Have you a spare compass to lend me?
D	B	G	I can spare you a compass.
D	B	H	I have no compasses but those in use.
D	B	J	I will send for the compass.
D	B	K	I will send you the compass.
D	B	L	You had better take a compass in the boat.
D	B	M	I will take a compass in the boat.
D	B	N	Our compasses do not agree.
D	B	P	Your compass must be out of the way.
D	B	Q	My compass is out of the way.

COURSE.

D	B	R	What do you think the best course to steer during the night (or the fog)?
D	B	S	What course shall I steer?

COURSE.—Continued.

D	B	T	What course will you steer?
D	B	V	I will steer ———.
D	B	W	You will steer the course indicated by signal.
D	C	B	NORTH.
D	C	F	North one quarter east.
D	C	G	North one half east.
D	C	H	North three quarters east.
D	C	J	North by east.
D	C	K	North by east one quarter east.
D	C	L	North by east one half east.
D	C	M	North by east three quarters east.
D	C	N	North north-east.
D	C	P	North north-east one quarter east.
D	C	Q	North north-east one half east.
D	C	R	North north-east three quarters east.
D	C	S	North-east by north.
D	C	T	North-east three quarters north.
D	C	V	North-east half north.
D	C	W	North-east one quarter north.
D	F	B	North-east.
D	F	C	North-east one quarter east.
D	F	G	North-east half east.
D	F	H	North-east three quarters east.
D	F	J	North-east by east.
D	F	K	North-east by east one quarter east.
D	F	L	North-east by east half east.

Yachting Code of Signals.

COURSE.—Continued.

D	F	M	North-east by east three quarters east.
D	F	N	East north-east.
D	F	P	East by north three quarters north.
D	F	Q	East by north half north.
D	F	R	East by north one quarter north.
D	F	S	East by north.
D	F	T	East three quarters north.
D	F	V	East half north.
D	F	W	East one quarter north.
D	G	B	EAST.
D	G	C	East one quarter south.
D	G	F	East one half south.
D	G	H	East three quarters south.
D	G	J	East by south.
D	G	K	East by south one quarter south.
D	G	L	East by south one half south.
D	G	M	East by south three quarters south.
D	G	N	East south-east.
D	G	P	South-east by east three quarters east.
D	G	Q	South-east by east half east.
D	G	R	South-east by east one quarter east.
D	G	S	South-east by east.
D	G	T	South-east three quarters east.
D	G	V	South-east half east.
D	G	W	South-east one quarter east.
D	H	B	South-east.

COURSE.—Continued.

D	H	C	South-east one quarter south.
D	H	F	South-east half south.
D	H	G	South-east three quarters south.
D	H	J	South-east by south.
D	H	K	South-east by south one quarter south.
D	H	L	South-east by south half south.
D	H	M	South-east by south three quarters south.
D	H	N	South south-east.
D	H	P	South by east three quarters east.
D	H	Q	South by east half east.
D	H	R	South by east one quarter east.
D	H	S	South by east.
D	H	T	South three quarters east.
D	H	V	South half east.
D	H	W	South one quarter east.
D	J	B	SOUTH.
D	J	C	South one quarter west.
D	J	F	South half west.
D	J	G	South three quarters west.
D	J	H	South by west.
D	J	K	South by west one quarter west.
D	J	L	South by west half west.
D	J	M	South by west three quarters west.
D	J	N	South south-west.
D	J	P	South-west by south three quarters south.
D	J	Q	South-west by south half south.

Yachting Code of Signals.

COURSE.—Continued.

D	J	R	South-west by south one quarter south.
D	J	S	South-west by south.
D	J	T	South-west three quarters south.
D	J	V	South-west half south.
D	J	W	South-west one quarter south.
D	K	B	**South-west.**
D	K	C	South-west one quarter **west**.
D	K	F	South-west half **west**.
D	K	G	South-west three quarters west.
D	K	H	South-west by west.
D	K	J	South-west by west one quarter west.
D	K	L	South-west by west half west.
D	K	M	South-west by west three quarters west.
D	K	N	West south-west.
D	K	P	West by south three quarters south.
D	K	Q	**West** by south half south.
D	K	R	West by south one quarter south.
D	K	S	West by south.
D	K	T	West three quarters south.
D	K	V	West half south.
D	K	W	West one quarter south.
D	L	B	WEST.
D	L	C	West one quarter north.
D	L	F	West half **north**.
D	L	G	West three quarters north.
D	L	H	West by north.

Yachting Code of Signals.

COURSE.—Continued.

D	L	J	West by north one quarter north.
D	L	K	West by north half north.
D	L	M	West by north three quarters north.
D	L	N	West north-west.
D	L	P	North-west by west three quarters west.
D	L	Q	North-west by west half west.
D	L	R	North-west by west one quarter west.
D	L	S	North-west by west.
D	L	T	North-west three quarters west.
D	L	V	North-west half west.
D	L	W	North-west one quarter west.
D	M	B	North-west.
D	M	C	North-west one quarter north.
D	M	F	North-west half north.
D	M	G	North-west three quarters north.
D	M	H	North-west by north.
D	M	J	North-west by north one quarter north.
D	M	K	North-west by north half north.
D	M	L	North-west by north three-quarters north.
D	M	N	North north-west.
D	M	P	North by west three quarters west.
D	M	Q	North by west half west.
D	M	R	North by west one quarter west.
D	M	S	North by west.
D	M	T	North three quarters west.
D	M	V	North half west.
D	M	W	North one quarter west.

Yachting Code of Signals.

CURRENT.

D	N	B	There is a northerly current.
D	N	C	There is a southerly current.
D	N	F	There is an easterly current.
D	N	G	There is a westerly current.
D	N	H	What current is there?

D

DAMAGE.

D	N	J	Have you sustained any damage?
D	N	K	Did the other vessel sustain damage?
D	N	L	We have sustained no damage?
D	N	M	Our damage is very slight.
D	N	P	Our damage is serious.
D	N	Q	The damage to the other vessel was slight.
D	N	R	The damage to the other vessel was serious.
D	N	S	The other vessel sustained no damage.

DAMAGES.

D	N	T	Have you a claim for damages?
D	N	V	When were the damages sustained?

DAMAGES.—Continued.

D	N	W	How were the damages sustained?
D	P	B	You will have a claim for damages.
D	P	C	We have a claim for damages.
D	P	F	A claim for damages is pending.
D	P	G	The damages were occasioned by collision.
D	P	H	The damages occurred on ———.

DANCE.

D	P	J	When will the dance come off?
D	P	K	There will be a dance on shore to-night.
D	P	L	There will be a dance on board the ———.
D	P	M	There is a dance on board the ———.
D	P	N	I am going to the dance on shore.
D	P	Q	I am going to the dance on board the ———.
D	P	R	I am not going to the dance.
D	P	S	We will have a dance if you will come on board.

DANCING.

D	P	T	Will there be dancing?
D	P	V	There will be dancing.
D	P	W	There will not be dancing.

DANGER.

D	Q	B	I fear danger.
D	Q	C	There is no danger to be feared.
D	Q	F	There is danger ahead.
D	Q	G	There is danger to leeward.
D	Q	H	If you stand on, there is danger of going ashore.
D	Q	J	There is no danger, you may stand on.
D	Q	K	Do not be too sanguine of there being no danger.

DANGEROUS.

D	Q	L	Is the coast dangerous?
D	Q	M	Is the harbor dangerous?
D	Q	N	The coast is dangerous, do not depend upon the soundings.
D	Q	P	The coast is not dangerous; shore is bold, but soundings are regular.
D	Q	R	The harbor is dangerous.
D	Q	S	The harbor is not dangerous.
D	Q	T	You will find it dangerous, be careful.

DARK.

D	Q	V	May I anchor at dark?
D	Q	W	Shall we anchor before dark?
D	R	B	Shall we anchor after dark?
D	R	C	Will it answer to run after dark?
D	R	F	I shall anchor before dark.
D	R	G	I shall not anchor until after dark.

DARK.—Continued.

D	R	H	Before dark.
D	R	J	After dark.
D	R	K	The night was very dark.
D	R	L	You must be careful running after dark.
D	R	M	At dark we shall heave to.
D	R	N	If the night is dark, anchor without further orders.
D	R	P	I do not wish to run after dark.
D	R	Q	The night promises to be very dark.
D	R	S	I do not think the night will be dark.
D	R	T	A dark night is the best.
D	R	V	A dark night will not answer.

DATE.

D	R	W	What is the latest date you have from New York?
D	S	B	Do they bring any later dates?
D	S	C	My latest date from New York is ———.
D	S	F	They have brought later dates.
D	S	G	They have no dates later than last received.
D	S	H	From New York.
D	S	J	From Boston.
D	S	K	From Philadelphia.
D	S	L	From Baltimore.
D	S	M	From Washington.
D	S	N	The first.
D	S	P	The second.

Yachting Code of Signals.

DATES.—Continued.

D	S	Q	The third.
D	S	R	The fourth.
D	S	T	The fifth.
D	S	V	The sixth.
D	S	W	The seventh.
D	T	B	The eighth.
D	T	C	The ninth.
D	T	F	The tenth.
D	T	G	The eleventh.
D	T	H	The twelfth.
D	T	J	The thirteenth.
D	T	K	The fourteenth.
D	T	L	The fifteenth.
D	T	M	The sixteenth.
D	T	N	The seventeenth.
D	T	P	The eighteenth.
D	T	Q	The nineteenth.
D	T	R	The twentieth.
D	T	S	The twenty-first.
D	T	V	The twenty-second.
D	T	W	The twenty-third.
D	V	B	The twenty-fourth.
D	V	C	The twenty-fifth.
D	V	F	The twenty-sixth.
D	V	G	The twenty-seventh.
D	V	H	The twenty-eighth.

DATES.—Continued.

D	V	J	The twenty-ninth.
D	V	K	The thirtieth.
D	V	L	The thirty-first.

DAY.

D	V	M	On what day?
D	V	N	What kind of a day did they have?
D	V	P	To-day.
D	V	Q	The next day.
D	V	R	The day before.
D	V	S	The day following.
D	V	T	To-morrow.
D	V	W	Day after to-morrow.
D	W	B	Yesterday.
D	W	C	Day before yesterday.
D	W	F	Monday.
D	W	G	Tuesday.
D	W	H	Wednesday.
D	W	J	Thursday.
D	W	K	Friday.
D	W	L	Saturday.
D	W	M	Sunday.
D	W	N	If the day is fine.
D	W	P	I think it will be a fine day.
D	W	Q	I think it will be a bad day.

Yachting Code of Signals.

DAY.—Continued.

D	W	R	The day was all that could be desired.
D	W	S	The day was unfavorable.

DELAY—ED.

D	W	T	What was the reason of the delay?
D	W	V	Why do you delay?
F	B	C	Do you delay your departure?
F	B	D	Will you be delayed?
F	B	G	Were the orders delayed?
F	B	H	Do not delay.
F	B	J	There will be no delay.
F	B	K	The delay is (was) unavoidable.
F	B	L	The delay is unnecessary.
F	B	M	I shall delay departure.
F	B	N	I cannot delay departure.
F	B	P	Send an answer without delay.
F	B	Q	I was delayed unexpectedly.
F	B	R	We shall be delayed.
F	B	S	We shall not be delayed.
F	B	T	The orders were delayed.
F	B	V	They must not be delayed.

DINE—D.

F	B	W	Will you come and dine with me?
F	C	B	Will you and your friends dine with me?

DINE—D.—Continued.

F	C	D	At what hour do you dine?
F	C	G	Do you dine ashore?
F	C	H	Do you dine on board the ———?
F	C	J	Will they dine with you?
F	C	K	Have you dined?
F	C	L	I shall be happy to dine with you.
F	C	M	It will be impossible for me to dine with you.
F	C	N	We dine at ———.
F	C	P	We expect to dine ashore.
F	C	Q	We shall not dine ashore.
F	C	R	We dine on board the ———.
F	C	S	I am going to dine with you.
F	C	T	He will dine with me.
F	C	V	He will not dine with me.
F	C	W	Thank you, I have dined.

DINING.

F	D	B	Is he dining ashore?
F	D	C	He is dining ashore.
F	D	G	He is dining on board the ———.
F	D	H	He is dining alone.

DINNER.

F	D	J	Do you attend the dinner ashore?
F	D	K	Do you attend the dinner on board the ———?

Yachting Code of Signals.

DINNER.—Continued.

F	D	L	Are you invited to the dinner?
F	D	M	By whom is the dinner given?
F	D	N	I shall attend the dinner ashore.
F	D	P	I shall attend the dinner on board the ———.
F	D	Q	I shall not attend the dinner ashore.
F	D	R	I shall not attend the dinner on board the ———.
F	D	S	I have no invitation to the dinner.
F	D	T	I am invited to the dinner.
F	D	V	Dinner is ready.
F	D	W	I will wait dinner.
F	G	B	The dinner need not wait for them.
F	G	C	I cannot wait dinner.

DRESS—ED.

F	G	D	Is it to be a full dress affair?
F	G	H	Do you go in full dress?
F	G	J	Shall we dress ship?
F	G	K	I shall go in full dress.
F	G	L	It will be a full dress affair.
F	G	M	Full dress is not expected.
F	G	N	Will you wear a dress coat?
F	G	P	Dress coats will not be expected.
F	G	Q	The yachts will be dressed.
F	G	R	The yachts will not be dressed.
F	G	S	Prepare to dress ship on signal.

DRESS—ED.—Continued.

F	G	T	Dress ship.
F	G	V	Will the yacht be dressed?
F	G	W	Will the squadron be dressed?
F	H	B	The squadron will be dressed.
F	H	C	The squadron will not be dressed.
F	H	D	The yacht will be dressed.
F	H	G	The yacht will not be dressed.
F	H	J	The squadron is (was) beautifully dressed.

DRY.

F	H	K	At what stage of the tide is the bank dry?
F	H	L	The bank is dry at low water.
F	H	M	The bank is dry at half tide.
F	H	N	We are high and dry at low water.
F	H	P	We left her high and dry.
F	H	Q	She is high and dry.
F	H	R	She ran up high and dry.

DURING.

F	H	S	During the day.
F	H	T	During the evening.
F	H	V	During the night.
F	H	W	During the week.
F	J	B	During the month.

E

EACH.

F	J	C	Each other.
F	J	D	Each of us.
F	J	G	Each of you.
F	J	H	Each of them.
F	J	K	Each of the ———.

EAST.

F	J	L	Off the east end of ———.
F	J	M	Off the east side of ———.
F	J	N	You will steer east.
F	J	P	I am steering east.
F	J	Q	I have been steering east.
F	J	R	Easterly.

EASTWARD.

F	J	S	Have you had the wind from the eastward?
F	J	T	Is anything in sight to the eastward?
F	J	V	Shall I stand to the eastward?
F	J	W	Shall I pass to the eastward?
F	K	B	I have had the wind from the eastward.
F	K	C	We shall probably have the wind from the eastward.
F	K	D	There is nothing in sight to the eastward.

EASTWARD.—Continued.

F	K	G	The squadron is in sight to the eastward.
F	K	H	There is a yacht in sight to the eastward.
F	K	J	Stand to the eastward.
F	K	L	Pass to the eastward.
F	K	M	The yachts are standing to the eastward.

ENDANGER.

F	K	N	You will endanger your spars.
F	K	P	You will endanger your rigging.
F	K	Q	You will endanger your vessel.

ENDEAVOR—ED.

F	K	R	Will you endeavor to do so?
F	K	S	I will endeavor to do so.
F	K	T	I endeavored to do so.
F	K	V	He endeavored to do so.
F	K	W	He will endeavor to do so.

ENFORCE—D.

F	L	B	Enforce the order.
F	L	C	I will enforce the order.
F	L	D	I cannot enforce the order.
F	L	G	The order (orders) must be enforced.

Yachting Code of Signals.

ENFORCE—D.—Continued.

F	L	H	The order has been enforced.
F	L	J	You will see the order enforced.

ENGAGED.

F	L	K	Are you engaged at present?
F	L	M	Will you be engaged?
F	L	N	I am engaged.
F	L	P	I am not engaged.
F	L	Q	If you are not engaged.
F	L	R	I shall be engaged.
F	L	S	I shall not be engaged.

ENGAGEMENT.

F	L	T	I hope you have no previous engagement.
F	L	V	I have no previous engagement.
F	L	W	I have a previous engagement.
F	M	B	A previous engagement will prevent me.

ENOUGH.

F	M	C	Have you enough?
F	M	D	I have enough.
F	M	G	I have not enough.
F	M	H	You will have wind enough.

ENOUGH.—Continued.

F	M	J	There will not be wind enough.
F	M	K	She carries canvas enough.
F	M	L	She has not canvas enough.

ENSIGN.

F	M	N	Is her ensign at half mast?
F	M	P	Is her ensign Union down?
F	M	Q	Can you distinguish her ensign?
F	M	R	Shall we set the ensign?
F	M	S	What ensign does she show?
F	M	T	Her ensign is at half mast.
F	M	V	Her ensign is Union down.
F	M	W	Set your ensign.
F	N	B	Her ensign does not blow out clear.
F	N	C	See that your ensign blows out clear
F	N	D	Her ensign cannot be distinguished.

ENTER.

F	N	G	Shall I enter the harbor (river)?
F	N	H	Enter the harbor and report upon it.
F	N	J	Enter the harbor (river).
F	N	K	Do not enter the harbor (river).

ENTERTAIN.

F	N	L	How many will you entertain?
F	N	M	How many can you entertain?
F	N	P	Will you entertain the proposition?
F	N	Q	I cannot entertain more than ———.
F	N	R	I can entertain ———.
F	N	S	Shall be happy to entertain your friends.

ENTERTAINED.

F	N	T	How were you entertained?
F	N	V	We were handsomely entertained.
F	N	W	We were not well entertained.
F	P	B	The proposition will be entertained.
F	P	C	The proposition cannot be entertained.

ENTERTAINMENT.

F	P	D	Will you attend the entertainment?
F	P	G	I cannot attend the entertainment.
F	P	H	I will attend the entertainment.
F	P	J	The entertainment will be a brilliant affair.

ENTRANCE.

F	P	K	Is the entrance of the harbor (river) clear?
F	P	L	Shall I stand in to the entrance of the harbor (river)?

ENTRANCE.—Continued.

F	P	M	The entrance is clear of danger.
F	P	N	The entrance is dangerous, be careful.
F	P	Q	I know nothing of the entrance.
F	P	R	Stand in as far as the entrance of the harbor (river).

ERROR.

F	P	S	Was there not an error in the signal?
F	P	T	By whom was the error committed?
F	P	V	What was the error?
F	P	W	There was an error in the signal.
F	Q	B	There was no error in the signal.
F	Q	C	There has been a gross error committed.
F	Q	D	The error was unavoidable.
F	Q	G	The error must not be repeated.
F	Q	H	The error will not be repeated.
F	Q	J	There was an error in time.
F	Q	K	There was an error in the distance.

ESCAPE—D.

F	Q	L	How did you escape?
F	Q	M	Cannot I escape?
F	Q	N	It was a wonderful escape.
F	Q	P	I made a narrow escape.
F	Q	R	There is no escape.

ESCAPE—D.—Continued.

F	Q	S	You can escape.
F	Q	T	We narrowly escaped being beaten.
F	Q	V	She narrowly escaped being beaten.

EVENING.

F	Q	W	This evening.
F	R	B	Last evening.
F	R	C	To-morrow evening.
F	R	D	During the evening.
F	R	G	———— Evening.

EXCEL—LED.

F	R	H	Do we excel?
F	R	J	You certainly excel.
F	R	K	I cannot say that you excel.
F	R	L	You excelled yourself.
F	R	M	We think we excelled.

EXCELLENT.

F	R	N	I think it was excellent.
F	R	P	The manœuvring was excellent.
F	R	Q	The race was an excellent one.

Yachting Code of Signals.

EXECUTE—D.

F	R	S	Can I depend upon the order (commission) being promptly executed?
F	R	T	When do you wish the order executed?
F	R	V	Execute the order promptly.
F	R	W	I will execute the order (commission) promptly.
F	S	B	The order (exercise) was well executed.
F	S	C	The order (exercise) was badly executed.
F	S	D	The squadron evolutions were well executed.
F	S	G	The squadron evolutions were badly executed.
F	S	H	I have executed your order (commission).

F

FAST.

F	S	J	Is she very fast?
F	S	K	Did she make fast time?
F	S	L	Was the time fast?
F	S	M	Shall I make fast?
F	S	N	She is very fast.
F	S	P	She is not fast.
F	S	Q	She made fast time.
F	S	R	She did not make fast time.
F	S	T	The time was fast.
F	S	V	The time was not fast.

FAST.—Continued.

F S W   You will make fast.
F T B   You will not make fast.

FATHOMS.

F T C   How many fathoms water?
F T D   ———— fathoms.

FAVOR.

F T G   Will you do me a favor?
F T H   Shall be happy to do you a favor.
F T J   Favor me with a visit when convenient.
F T K   You will favor me very much.

FEAR.

F T L   Have you any fear?
F T M   I have fear.
F T N   I have no fear.
F T P   Have no fear of danger, the way is clear.
F T Q   I fear nothing and will stand on.
F T R   We fear danger and will heave to.
F T S   Not without fear.

FEARFUL.

F	T	V	There was a fearful sea running.
F	T	W	It is a fearful coast in bad weather.
F	V	B	There is (was) a fearful surf breaking.

FEEL.

F	V	C	Shall I feel my way in with the lead?
F	V	D	Do you feel like having a good time?
F	V	G	Do you feel like having dinner?
F	V	H	Do you feel sleepy?
F	V	J	Do you feel fatigued?
F	V	K	Feel your way in with the lead.
F	V	L	Do not attempt to feel your way in.
F	V	M	I feel like having a good time.
F	V	N	I feel very hungry.
F	V	P	I feel like having dinner.
F	V	Q	I am feeling sleepy and will turn in.
F	V	R	I feel much fatigued.
F	V	S	I do not feel fatigued.

FEELING.

F	V	T	Are you feeling well?
F	V	W	I am feeling well.
F	W	B	I am not feeling well.
F	W	C	You are totally without feeling for us.
F	W	D	I have the kindest feeling for you.

Yachting Code of Signals.

FEET.

F	W	G	In how many feet of water are you?
F	W	H	There are but a few feet of water under me.

FINE.

F	W	J	Did she present a fine appearance?
F	W	K	She presented a fine appearance.

FIREWORKS.

F	W	L	Is there to be a display of fireworks?
F	W	M	Will you have a display of fireworks?
F	W	N	There will be a display of fireworks.
F	W	P	We shall have a display of fireworks.

FISH.

F	W	Q	Can we catch fish in this vicinity?
F	W	R	Fish abound in these waters.

FISHING.

F	W	S	Will you go fishing with us?
F	W	T	How did you find the fishing?
F	W	V	I will go with you fishing.
G	B	C	I cannot go fishing.

FISHING.—Continued.

G	B	D	The fishing was (is) excellent.
G	B	F	I had poor luck fishing.
G	B	H	We are going fishing.

FISHING-LINES.

G	B	J	Have you any fishing-lines?
G	B	K	Have you any bait?
G	B	L	Shall we bring fishing-lines with us?
G	B	M	We have plenty of fishing-lines.
G	B	N	We have no fishing-lines.
G	B	P	Bring fishing-lines with you.

FISHERMEN.

G	B	Q	The fishermen can (will) inform you.
G	B	R	Ask assistance from the fishermen.
G	B	S	The fishermen were very clever.
G	B	T	The fishermen would give no information (assistance).
G	B	V	You will find fishermen there.

FOG.

G	B	W	We were in a dense fog.
G	C	B	In the fog to windward.
G	C	D	In the fog to leeward.

Yachting Code of Signals.

FOG.—Continued.

G	C	F	In the fog ahead.
G	C	H	In the fog astern.
G	C	J	The fog bank to windward is threatening.
G	C	K	On account of the fog.
G	C	L	When the fog lifts.

FRIGHTENED.

G	C	M	Were you frightened?
G	C	N	The fact is we were all frightened.

PROM.

G	C	P	Where are you from?
G	C	Q	Have you heard from ———?
G	C	R	When did you hear from ———?
G	C	S	I have heard from ———.
G	C	T	I have not heard from ———.
G	C	V	I am anxious to hear from ———.

FUTURE.

G	C	W	At some future period.
G	D	B	It may take place in future.
G	D	C	I may have better luck in future.
G	D	F	You will have better luck in future.

G

GAFF.

G	D	H	I have sprung my main gaff.
G	D	J	I have sprung my fore gaff.
G	D	K	My gaffs are both sprung.

GAIN.

G	D	L	Do you gain on her much?
G	D	M	What shall I gain by it?
G	D	N	We gain on her fast.
G	D	P	We gain on her a little.
G	D	Q	The yacht astern is gaining on us.
G	D	R	We are gaining on the yacht ahead.
G	D	S	You will gain nothing by it.
G	D	T	She has gained a great reputation for speed.

GLAD.

G	D	V	I was very glad to hear (know) it.
G	D	W	I am very glad to hear (know) it.
G	F	B	You will be glad to hear the news.

GO—ING.

G	F	C	Shall I go ahead?
G	F	D	Will you go?
G	F	H	Where are you going?
G	F	J	Are you going?
G	F	K	Are you going to ———?
G	F	L	When will you go?
G	F	M	I shall be glad to go.
G	F	N	I cannot go.
G	F	P	We are going.
G	F	Q	We are not going.
G	F	R	Go ahead.

GUESTS.

G	F	S	Whose guest is he?
G	F	T	He is my guest.
G	F	V	He is the guest of the squadron.
G	F	W	Shall be happy to receive your guests.

GUNS.

G	H	B	Shall I fire a gun?
G	H	C	Did you hear a gun fired?
G	H	D	In what direction do you hear the guns?
G	H	F	Shall I take a gun with me?
G	H	J	Fire a gun.

GUNS.—Continued.

G	H	K	Do not fire a gun.
G	H	L	We heard a gun to leeward.
G	H	M	We heard a gun to windward.
G	H	N	Signal guns are being fired.
G	H	P	Bring a gun with you.

H

HALLIARDS.

G	H	Q	Have you rope to spare for a pair of peak (throat) halliards?
G	H	R	I carried away my peak halliards.
G	H	S	I carried away my throat halliards.
G	H	T	I carried away my jib halliards.
G	H	V	I carried away my flying jib halliards.
G	H	W	I can spare you a new set of halliards.

HAPPEN.

G	J	B	How did it happen?
G	J	C	When did it happen?
G	J	D	Where did it happen?

Yachting Code of Signals.

HARBOR.

G	J	F	Can you enter the harbor?
G	J	H	Is any vessel acquainted with the harbor and willing to lead in?
G	J	K	Shall I make a harbor for the night or stand on?
G	J	L	What port or harbor shall I make to-night?
G	J	M	What harbor will you make to-night?
G	J	N	How long do you intend to remain in ———?
G	J	P	At what port or harbor do you propose to end the cruise?
G	J	Q	Will you make a harbor to-night?
G	J	R	Is the harbor a good one?
G	J	S	Enter the harbor and anchor.
G	J	T	Make the harbor to be designated.
G	J	V	Make any harbor you think proper.
G	J	W	Heave to off ——— and await further orders.
G	K	B	I shall remain in the harbor ——— days.
G	K	C	I shall remain in the harbor but a few hours.
G	K	D	I shall make a harbor to-night.
G	K	F	The harbor is good, depend upon the chart.
G	K	H	The harbor is dangerous and not well sheltered.

HEAD.

G	K	J	Are you not trimmed too much by the head?
G	K	L	Shall I head her off?
G	K	M	I am trimmed too much by the head.
G	K	N	Head her off if possible.

HEAR.

G	K	P	Did you hear (have you heard) anything of the ———?
G	K	Q	Did you hear the news?
G	K	R	I have heard nothing.
G	K	S	If you hear anything, let me know.
G	K	T	If I hear anything, I will let you know.

HOUR.

G	K	V	At what hour?
G	K	W	Has an hour been named?
G	L	B	Can you pass an hour with me?
G	L	C	At the hour appointed.
G	L	D	I am coming to pass an hour with you.
G	L	F	The hour has passed.
G	L	H	Twelve, Meridian.
G	L	J	One, P. M.
G	L	K	Two, "
G	L	M	Three, "
G	L	N	Four, "
G	L	P	Five, "
G	L	Q	Six, "
G	L	R	Seven, "
G	L	S	Eight, "
G	L	T	Nine, "
G	L	V	Ten, "
G	L	W	Eleven, "

Yachting Code of Signals.

HOUR.—Continued.

G	M	B	Twelve, Midnight.
G	M	C	One, A. M.
G	M	D	Two, "
G	M	F	Three, "
G	M	H	Four, "
G	M	J	Five, "
G	M	K	Six, "
G	M	L	Seven, "
G	M	N	Eight, "
G	M	P	Nine, "
G	M	Q	Ten, "
G	M	R	Eleven, "

HULL.

G	M	S	Is her hull damaged?
G	M	T	Hull down ahead.
G	M	V	Hull down astern.
G	M	W	Hull down to windward.
G	N	B	Hull down to leeward.
G	N	C	Her hull is damaged.
G	N	D	Her hull is not damaged.
G	N	F	Report the condition of her hull.

HURRY.

G	N	H	Are you in a hurry?
G	N	J	Hurry off, you have no time to lose.

I

IF.

G	N	K	If you will.
G	N	L	I will if you will.
G	N	M	If I can.
G	N	P	If you can.
G	N	Q	If you will not.
G	N	R	I will not if you cannot.
G	N	S	If you please.
G	N	T	If you wish.
G	N	V	If it is ordered.
G	N	W	If it is requested.
G	P	B	If it is possible.
G	P	C	If it is prudent.
G	P	D	If I am in time.

IMAGINE—D—ING.

G	P	F	What do you imagine?
G	P	H	I imagine nothing.
G	P	J	I imagine the worst.
G	P	K	I imagine the best.
G	P	L	So I imagined.
G	P	M	You are always imagining something.
G	P	N	I am always imagining something.

IMPERIL.

G	P	Q	Do you think we shall imperil the vessel?
G	P	R	You will imperil your safety, do not venture.
G	P	S	Imperil nothing.
G	P	T	I will imperil nothing.

IMPORTANCE.

G	P	V	Is it of importance?
G	P	W	It is of no importance.
G	Q	B	It is of great importance.

IMPORTANT.

G	Q	C	Is it important?
G	Q	D	Have you anything important for me?
G	Q	F	It is very important.
G	Q	H	It is not important.
G	Q	J	I have important information for you.
G	Q	K	I have nothing important for you.
G	Q	L	The news is of the most important character.
G	Q	M	The news is not important.

IMPROPER.

G	Q	N	Is it improper?
G	Q	P	Do you think it improper?
G	Q	R	It is very improper.

IMPROPER.—Continued.

G	Q	S	It is not improper.
G	Q	T	I think it improper.
G	Q	V	I do not think it improper.

IMPRUDENT.

G	Q	W	Do you think it imprudent?
G	R	B	I think it very imprudent.
G	R	C	I think it is not imprudent.
G	R	D	Do not be imprudent.
G	R	F	I will not be imprudent.

INABILITY.

G	R	H	I regret my inability.
G	R	J	By reason of your inability.

INFORM—ED.

G	R	K	Can you inform me?
G	R	L	Will you inform me?
G	R	M	Shall I inform him?
G	R	N	I will inform you.
G	R	P	I cannot inform you.
G	R	Q	If necessary, I will inform you.
G	R	S	If you hear anything, inform me.

Yachting Code of Signals.

INFORM—ED.—Continued.

G	R	T	You had better inform him.
G	R	V	You had better not inform him.
G	R	W	I will keep you informed.
G	S	B	Keep me informed.
G	S	C	You appear to be well informed.

INFORMATION.

G	S	D	Have you any information?
G	S	F	What information do you desire?
G	S	H	Do you require further information?
G	S	J	Was the information correct?
G	S	K	Who gave you the information?
G	S	L	I have no information.
G	S	M	I have important information.
G	S	N	I desire information.
G	S	P	I must wait for further information.
G	S	Q	No further information is required.
G	S	R	The information was correct.
G	S	T	The information was untrue.
G	S	V	Thanks for the information.
G	S	W	I will get further information if possible.
G	T	B	The information came from a good source.
G	T	C	The information must not be relied upon.
G	T	D	When I get the information.

INJURE.

G	T	F	Will it injure me?
G	T	H	Would you injure me?
G	T	J	Will it injure the vessel?
G	T	K	Will it injure the sails or rigging?
G	T	L	It will injure me.
G	T	M	It will not injure me.
G	T	N	I would not injure you.
G	T	P	It cannot injure you.
G	T	Q	It will injure nothing.

INJURY.

G	T	R	Have you sustained any injury?
G	T	S	We have sustained but slight injury.
G	T	V	We have sustained serious injury.
G	T	W	We have sustained no injury.
G	V	B	Be careful not to cause injury.
G	V	C	I will cause no injury.
G	V	D	The injury will be made good.

INJUSTICE.

G	V	F	You are doing yourself great injustice.
G	V	H	You do him great injustice.
G	V	J	You do the yacht great injustice.
G	V	K	I have done him injustice.

INJUSTICE.—Continued.

G	V	L	I have done him no injustice.
G	V	M	I have done the yacht injustice.
G	V	N	I have done him great injustice.

INQUIRE—D.

G	V	P	Shall I inquire?
G	V	Q	I wish you would inquire.
G	V	R	I will inquire.
G	V	S	Did you inquire?
G	V	T	I have inquired.
G	V	W	I have not inquired.

INSTANT—LY.

G	W	B	Do you wish it done instantly?
G	W	C	This instant.
G	W	D	At the instant.
G	W	F	Do it instantly.
G	W	H	It was done instantly.
G	W	J	They were off on the instant.
G	W	K	It required but an instant.

INSTRUCTIONS.

G	W	L	What are the instructions?
G	W	M	Will send you the instructions.

Yachting Code of Signals.

INSTRUCTIONS.—Continued.

G	W	N	The instructions have been sent.
G	W	P	The instructions will be sent.
G	W	Q	When will the instructions be sent?
G	W	R	Have you received the instructions?
G	W	S	The instructions have been received.
G	W	T	The instructions have not been received.

INVITATION.

G	W	V	Have you received an invitation?
H	B	C	Have you an invitation for me?
H	B	D	How many invitations are out?
H	B	F	Is an invitation necessary?
H	B	G	I have received an invitation.
H	B	J	I have not received an invitation.
H	B	K	An invitation has been sent you.
H	B	L	An invitation has been sent your friends.
H	B	M	I have an invitation for you.
H	B	N	I do not know how many invitations are out.
H	B	P	The invitations are numerous.
H	B	Q	The invitations are few.
H	B	R	The invitations have been limited.
H	B	S	Invitations are necessary.
H	B	T	Invitations are unnecessary.
H	B	V	The invitation came too late.
H	B	W	The invitation was in time.

Yachting Code of Signals.

INVITE.

H	C	B	Whom do you intend to invite?
H	C	D	Have you been invited?
H	C	F	How many have been invited?
H	C	G	I shall invite no more than I can accommodate.
H	C	J	I shall invite but few.
H	C	K	I have not been invited.
H	C	L	I do not expect to be invited.
H	C	M	I have been invited.
H	C	N	I do not know how many have been invited.

IT.

H	C	P	Where is it?
H	C	Q	How is it?
H	C	R	It is.
H	C	S	It was.
H	C	T	It will.

J

JOIN.

H	C	V	Will you join me?
H	C	W	Will you join the squadron?
H	D	B	When do you join the squadron?

JOIN.—Continued.

H	D	C	Are you ordered to join the squadron?
H	D	F	I will join you.
H	D	G	I cannot join you.
H	D	J	I shall be glad if you will join the party.
H	D	K	The party insist that you join them.
H	D	L	I will join the squadron.
H	D	M	I shall not join the squadron.
H	D	N	I am ordered to join the squadron.

JOKE—ING.

H	D	P	Was it a joke?
H	D	Q	Are you fond of a joke?
H	D	R	How will he stand a practical joke?
H	D	S	Were you joking?
H	D	T	It was a joke.
H	D	V	It was not a joke.
H	D	W	It was a serious joke for me.
H	F	B	Leave off joking, be serious.
H	F	C	I am not fond of jokes.
H	F	D	We will leave off joking.
H	F	G	We played a practical joke upon him.
H	F	J	Be careful, he will not stand a practical joke.
H	F	K	I was only joking.
H	F	L	I am not joking.
H	F	M	The joke was a good one.

JOKE—ING.—Continued.

H	F	N	The joke was not good.
H	F	P	The joke was understood and appreciated.

JUDGES.

H	F	Q	Did the judges decide fairly?
H	F	R	When will the judges give a final decision.
H	F	S	Are the judges appointed?
H	F	T	No judges have been appointed.
H	F	V	The judges decided fairly.
H	F	W	I do not think the judges decided fairly.
H	G	B	The decision of the judges is final.
H	G	C	I do not know when the judges will decide.

K

KEDGE.

H	G	D	Can you kedge her off?
H	G	F	Can you lend me a kedge anchor?
H	G	J	She can be got off without a kedge anchor.
H	G	K	Get your kedge anchor out astern.
H	G	L	I can get her off with a kedge anchor.
H	G	M	I will send you a kedge anchor.
H	G	N	The kedge anchor is not heavy enough.

KEEP.

H	G	P	Shall I keep company with you?
H	G	Q	Keep off more.
H	G	R	Keep your sails full.
H	G	S	Keep company with me.
H	G	T	Keep company with the squadron.
H	G	V	Keep in with the land.
H	G	W	Keep clear of the land.
H	J	B	Keep your lead going.
H	J	C	Keep a good lookout.
H	J	D	Keep him on board.
H	J	F	I will keep him on board.
H	J	G	I cannot keep him on board.

KIND—NESS.

H	J	K	Will you promise to be kind to me?
H	J	L	I will promise to be very kind to you.
H	J	M	You are (were) very kind.
H	J	N	I appreciate your kindness.
H	J	P	It was a kind act.

KNOW—N.

H	J	Q	Do you know it?
H	J	R	How do you (they) know it?
H	J	S	When will you know?

KNOW—N.—Continued.

H	J	T	Do you know it to be so?
H	J	V	Is it known?
H	J	W	When will it be known?
H	K	B	Was it known before?
H	K	C	You do not know it.
H	K	D	I know it to be true.
H	K	F	Let me know.
H	K	G	I will let you know.
H	K	J	I cannot let you know.
H	K	L	It is known.
H	K	M	It is not known.
H	K	N	It will never be known.
H	K	P	It will soon be known.
H	K	Q	I hope you will make it known.
H	K	R	I know.
H	K	S	I do not know.

L

LAND.—(See "Sight.")

H	K	T	May I run in and make the land?
H	K	V	Is the land in sight?
H	K	W	How does the land appear?

LAND.—Continued.—(See "Sight.")

H	L	B	Shall I keep the land in sight.
H	L	C	The land is not yet in sight.
H	L	D	The land is in sight ahead.
H	L	F	The land is in sight to leeward.
H	L	G	The land is in sight to windward.
H	L	J	The land is close aboard.
H	L	K	The land is high.
H	L	M	The land is low.
H	L	N	Keep the land in sight.
H	L	P	Run in and make the land.
H	L	Q	Stand off the land.
H	L	R	Stand in for the land.
H	L	S	Under the land.

LAST.

H	L	T	Where are you from last?
H	L	V	How long do you think it will last?
H	L	W	When were you last with us?
H	M	B	I am last from ———.
H	M	C	The last cruise.
H	M	D	Last year.
H	M	F	Last month.
H	M	G	Last week.
H	M	J	Last race.
H	M	K	I do not think it will last long.

LAST.—Continued.

H	M	L	It will last some time yet.
H	M	N	No saying how long it will last.
H	M	P	When I was with you last.

LATE.

H	M	Q	Am I late?
H	M	R	Why were you so late?
H	M	S	Will he be late in getting under weigh?
H	M	T	You are very late.
H	M	V	You are not late.
H	M	W	Our being late was unavoidable.
H	N	B	We are excusable for being late.
H	N	C	He will be late in getting under weigh.
H	N	D	Do not be late.
H	N	F	I shall not be late.
H	N	G	I shall be late.

LEAD.

H	N	J	Shall I heave the lead?
H	N	K	Heave the lead.
H	N	L	Keep heaving the lead.
H	N	M	I am heaving the lead.
H	N	P	You need not use the lead, the water is deep.

LEAK.

H	N	Q	I have sprung a leak.
H	N	R	The leak is serious.
H	N	S	The leak is small.
H	N	T	I am trying to stop the leak.
H	N	V	My boats leak badly.
H	N	W	The leak is so bad, I shall have to go into port.
H	P	B	I can stop the leak.
H	P	C	The leak cannot be stopped.
H	P	D	I do not know where the leak is.
H	P	F	Is the leak bad?
H	P	G	Cannot you stop the leak?

LEAST.

H	P	J	What is the least water you have had.
H	P	K	What is the least water we shall find?
H	P	L	The least water we have had is ———.
H	P	M	The least water you will find is ———.

LEAVE—ING.

H	P	N	When do you leave?
H	P	Q	Do you expect to leave soon?
H	P	R	When does the leave expire?
H	P	S	When will the squadron be leaving?
H	P	T	I am about to leave.

Yachting Code of Signals.

LEAVE—ING.—Continued.

H	P	V	I shall not leave for some time.
H	P	W	He is off on leave.
H	Q	B	His leave will expire soon.
H	Q	C	His leave will not expire for some time.
H	Q	D	When you are about to leave.
H	Q	F	When we are about to leave.
H	Q	G	I will leave when you say so.
H	Q	J	Do not leave me (us) alone.
H	Q	K	The squadron will be leaving shortly.
H	Q	L	The squadron will not leave for some time.
H	Q	M	The yacht is about leaving.

LEND.

H	Q	N	Will you lend me assistance?
H	Q	P	Can you lend me a boat?
H	Q	R	Can you lend me an anchor?
H	Q	S	Can you lend me a tow-line?
H	Q	T	Can you lend me a lead and line?
H	Q	V	Can you lend me a man (men)?
H	Q	W	Can you lend me a rope?
H	R	B	Shall I lend him assistance?
H	R	C	I will lend you anything you require.
H	R	D	Lend me assistance.

LIABLE.

H	R	F	Will I be liable?
H	R	G	Who is liable?
H	R	J	You will be liable.
H	R	K	They will be liable.
H	R	L	You will not be liable.
H	R	M	They will not be liable.
H	R	N	I am not liable.
H	R	P	I shall be liable.

LIABILITY.

H	R	Q	Is there any liability?
H	R	S	Where or with whom does the liability rest?
H	R	T	Who will assume the liability?
H	R	V	Will you assume the liability?
H	R	W	There will be no liability.
H	S	B	There will be liability.
H	S	C	I do not know where or with whom the liability will rest.
H	S	D	I do not know who will assume the liability.
H	S	F	The liability rests with ———.
H	S	G	He will assume the liability.
H	S	J	I will assume the liability.

LIKE—LY.

H	S	K	What would you like?
H	S	L	How did you like him?

Yachting Code of Signals.

LIKE—LY.—Continued.

H	S	M	Whenever you like.
H	S	N	I like him very much.
H	S	P	I should like to have you with me.
H	S	Q	It is very like her.
H	S	R	We should like to be with you.
H	S	T	One is very much like the other.
H	S	V	Is it likely?
H	S	W	It is very likely.
H	T	B	It is not very likely.

LINE.

H	T	C	Shall I give you a tow-line?
H	T	D	Do you want a tow-line?
H	T	F	Come alongside and give me a line.
H	T	G	Give her a tow-line.
H	T	J	The tow-line parted.
H	T	K	I am in want of a tow-line.

LITTLE.

H	T	L	Shall I stand on a little further?
H	T	M	Stand on a little further.
H	T	N	Keep a little more off.
H	T	P	Keep a little closer to the wind.
H	T	Q	You must do a little better than that.
H	T	R	I can do a little better.

LIVE.

H	T	S	Can a boat live in such a sea?
H	T	V	Can he live?
H	T	W	Where does he live?
H	V	B	No boat can live in such a sea.
H	V	C	He will live.
H	V	D	He cannot live.
H	V	F	He lives at ———.

LOSS.

H	V	G	Was the loss serious?
H	V	J	What loss have you sustained?
H	V	K	How was the loss occasioned?
H	V	L	The loss was serious.
H	V	M	The loss was small.
H	V	N	There was no loss.
H	V	P	I have sustained no loss.
H	V	Q	I am at a loss to understand.
H	V	R	It was a great loss to them.
H	V	S	I hope that you have sustained no loss.
H	V	T	The loss was occasioned by ———.

LOST.

H	V	W	Have you lost a man (men)?
H	W	B	What have you lost?

LOST.—Continued.

H	W	C	How was it lost.
H	W	D	When was it lost?
H	W	F	Why have you lost time?
H	W	G	I have lost a man (men).
H	W	J	I have lost time.
H	W	K	You have lost time.
H	W	L	No time has been lost.
H	W	M	No time will be lost.
H	W	N	It was lost by carelessness.
H	W	P	Lost overboard.
H	W	Q	It was lost.

LUCK—Y.

H	W	R	You are in luck.
H	W	S	I am in luck.
H	W	T	The luck appears to have changed.
H	W	V	The luck is always one way.
J	B	C	I hope the luck will change.
J	B	D	The luck will change.
J	B	F	It is (was) just my luck.
J	B	G	It appears to be just your luck.
J	B	H	Who is (was) the lucky one?
J	B	K	He is lucky.
J	B	L	You are always lucky.

LUNCH.

J	B	M	Will you join us at lunch?
J	B	N	At what hour do you lunch?
J	B	P	Is lunch ready?
J	B	Q	Will you lunch on board the ———?
J	B	R	Shall we take a lunch with us?
J	B	S	I will join you at lunch.
J	B	T	I shall lunch on board the ———.
J	B	V	Lunch is ready.
J	B	W	You had better take a lunch with you.
J	C	B	Bring the lunch basket.

LURCH.

J	C	D	In a heavy lee lurch.
J	C	F	In a heavy lurch to windward.

M

MADE.

J	C	G	When was the land made?
J	C	H	I made the land on the ———.
J	C	K	You have made a fine passage.
J	C	L	I have made a fine passage.

MADE.—Continued.

C	M	We have made a fine day's run.
C	N	I have made the land.
C	P	When I made the land.

MAKE.

C	Q	How much do you make it?
C	R	Do you expect to make the land soon?
C	S	What do you make it out to be?
C	T	Shall I make a harbor (port)?
C	V	When you make the land report it.
C	W	We should make the land soon.
D	B	We should make the land ahead.
D	C	We should make the land on the lee bow.
D	F	We should make the land on the weather bow.
D	G	Make the best of your way to ———.
D	H	I hope you will make a quick passage.
D	K	I hope to make a quick passage.
D	L	Make a note of it.
D	M	Make any port (harbor) you can.
D	N	Make provision for any emergency.

MANY.

D	P	How many?
D	Q	Have you seen many?

MANY.—Continued.

J	D	R	Will many be there?
J	D	S	Not many.
J	D	T	I have not seen many.
J	D	V	I have seen a great many.
J	D	W	A great many will be there.
J	F	B	Many days.

MAST.

J	F	C	Mainmast.
J	F	D	Foremast.
J	F	G	Maintopmast.
J	F	H	Foretopmast.
J	F	K	I have sprung my mainmast.
J	F	L	I have sprung my maintopmast.
J	F	M	I have sprung my foremast.
J	F	N	I have sprung my foretopmast.
J	F	P	Have you sprung your masts?
J	F	Q	My masts are both sprung.
J	F	R	Have you a spare topmast?
J	F	S	I have a spare topmast.
J	F	T	Has she carried away a mast?
J	F	V	She has carried away a mast.
J	F	W	At half mast.
J	G	B	Will have to fish my foremast.
J	G	C	Will have to fish my mainmast.

Yachting Code of Signals.

MATTER.

G	D	What is the matter?
G	F	Is anything the matter?
G	H	Will you take the matter into consideration?
G	K	There is nothing the matter.
G	L	It is no matter.
G	M	The matter is important.
G	N	It is an insignificant matter.
G	P	The matter is not worth consideration.

MEDICINE.

G	Q	Do you require medicine?
G	R	I require some medicine.
G	S	My medicine chest is at your service.
G	T	I will send for your medicine chest.
G	V	I will send you my medicine chest.

MEN.

G	W	What number of men have you?
H	B	Do you want any men?
H	C	Am short of men, can you spare any?
H	D	Are your men all aboard?
H	F	What kind of men have you?
H	G	Send at once any men you can spare.
H	K	Will do so.

MEN.—Continued.

J	H	L	Some of my men are ashore.
J	H	M	When your men are aboard report it.
J	H	N	My men are not good for much.
J	H	P	My men are very good.
J	H	Q	My men are exhausted.
J	H	R	My men are all aboard.

MESSAGE.

J	H	S	Have you any message for me?
J	H	T	Will you send a message by me?
J	H	V	I have a message for you.
J	H	W	I will send a message by you.

MET.

J	K	B	Have you met anything on the passage?
J	K	C	Have you met with a mishap?
J	K	D	I have met with nothing on the passage.
J	K	F	I met with ——— on the passage.
J	K	G	I have met with a mishap.

MINUTES.

J	K	H	How many minutes?
J	K	L	——— minutes.

Yachting Code of Signals.

MINUTES.—Continued.

K	M	In a few minutes.
K	N	You have not a minute to spare.
K	P	Wait a few minutes.

MISS—ED.

K	Q	Did you miss me?
K	R	Did you miss stays?
K	S	I missed you very much.
K	T	We missed stays.

MISTAKE.

K	V	Was it a mistake?
K	W	Did you not make a mistake?
L	B	How did the mistake occur?
L	C	It was a mistake.
L	D	You have made a mistake.
L	F	A mistake has been made.
L	G	The mistake was unavoidable.
L	H	The mistake was caused by carelessness.

MONTH.

L	K	In what month?
L	M	During what month (months)?

MONTH.—Continued.

J	L	N	During the month (months) of ———.
J	L	P	This month.
J	L	Q	Next month.
J	L	R	Last month.
J	L	S	January.
J	L	T	February.
J	L	V	March.
J	L	W	April.
J	M	B	May.
J	M	C	June.
J	M	D	July.
J	M	F	August.
J	M	G	September.
J	M	H	October.
J	M	K	November.
J	M	L	December.

MUCH.

J	M	N	How much.
J	M	P	Not much.
J	M	Q	Very much.

Yachting Code of Signals.

N

NAME.

M	R	What name?
M	S	What is her name?
M	T	The name is (was) not understood.
M	V	The name is (was) understood.

NATIONS.

M	W	United States.
N	B	England or English.
N	C	France or French.
N	D	North Germany or German.
N	F	Italy or Italian.
N	G	Spain or Spanish.
N	H	Denmark or Danish.
N	K	Sweden or Swedish.
N	L	Norway or Norwegian.
N	M	Russia or Russian.
N	P	Belgium or Belgian.
N	Q	Holland.
N	R	Turkey or Turkish.
N	S	Brazil or Brazilian.
N	T	Mexico or Mexican.
N	V	Peru or Peruvian.
N	W	Chili or Chilian.

NEAR.

J	P	B	How near?
J	P	C	Very near.
J	P	D	Too near.
J	P	F	You must be near the land.
J	P	G	We are near the land.
J	P	H	The land is near.

NEW.

J	P	K	Is there anything new?
J	P	L	There is nothing new.

NEWS.

J	P	M	What is the news?
J	P	N	What is the latest news from New York?
J	P	Q	There is no news.
J	P	R	The news is of importance.
J	P	S	The news is of no importance.
J	P	T	I have news for you.
J	P	V	I have no news for you.

NIGHT.

P	W	At night.
Q	B	During the night.
Q	C	Last night.
Q	D	To-night.
Q	F	To-morrow night.
Q	G	——— night.
Q	H	Was it a dark night?
Q	K	It was a very dark night.
Q	L	The night was not dark.

NORTH—ERLY.

Q	M	Have you had a northerly wind?
Q	N	North of ———.
Q	P	On the north side or end of ———.
Q	R	On the north shore.
Q	S	The wind was northerly.
Q	T	We have had a strong northerly current.
Q	V	On a northerly course.
Q	W	In a northerly direction.
R	B	We have had northerly winds.
R	C	Is anything in sight to the northward?
R	D	Shall I stand to the northward?
R	F	Shall I pass to the northward?
R	G	Northerly.

Yachting Code of Signals.

NUMERALS.

(To be made **without** the Club Flag.)

B	One,	1.
F	Two,	2.
G	Three,	3.
H	Four,	4.
J	Five,	5.
K	Six,	6.
L	Seven,	7.
P	Eight,	8.
Q	Nine,	9.
R	Naught,	o (cypher).
T	Hundred,	oo.
W	Thousand,	ooo.

OF.

J	R	H	Of course.
J	R	K	Of course not.

OFF.

J	R	L	When will you be off?
J	R	M	How far off?

Yachting Code of Signals.

OFF.—Continued.

R	N	Is he off?
R	P	Shall we be off?
R	Q	Not far off.
R	S	Very far off.
R	T	Off the coast of ———.
R	V	Off shore.
R	W	Off the wind.
S	B	He is just off.
S	C	He is off at last.
S	D	Stand off the land.
S	F	Stand off and on shore (harbor).
S	G	Come, let us be off.
S	H	Off.

OFTEN.

S	K	How often?
S	L	Does it occur often?
S	M	Very often.
S	N	Not often.
S	P	It occurs often.
S	Q	It does not occur often.
S	R	It occurs too often.

OPINION.

J	S	T	What is your opinion?
J	S	V	Have you expressed an opinion?
J	S	W	Was an opinion expressed?
J	T	B	Who expressed the opinion?
J	T	C	I have no opinion.
J	T	D	The opinion is good.
J	T	F	The opinion is bad.
J	T	G	No opinion was expressed.
J	T	H	A decided opinion was expressed.
J	T	K	The opinion is unpopular.
J	T	L	Such is the popular opinion.
J	T	M	It is the opinion of the senior officer.
J	T	N	His opinion should be taken.
J	T	P	I have taken his opinion.

OPPORTUNITY

J	T	Q	Is it a good opportunity?
J	T	R	When will the opportunity offer?
J	T	S	Take advantage of the opportunity.
J	T	V	The opportunity must not be neglected.
J	T	W	It is a good opportunity.
J	V	B	I had no opportunity.
J	V	C	There will be no opportunity for some time.
J	V	D	The opportunity will offer soon.
J	V	F	There is (will be) no opportunity.

ORDERS.

V	G	What orders have you received?
V	H	Have you received the orders?
V	K	When were the orders received?
V	L	Shall I obey the orders?
V	M	In what shape were the orders received?
V	N	Will you send orders?
V	P	The orders have been sent.
V	Q	The orders have been received.
V	R	The orders have not been received.
V	S	You will obey the orders.
V	T	Other orders will be sent.
V	W	No other orders will be sent.
W	B	The orders were in writing.
W	C	The orders were full.
W	D	The orders were verbal.
W	F	The orders will be signalled.
W	G	I will send orders.

OTHER.

W	H	What other?
W	K	Which other?
W	L	Have you any other?
W	M	Any other.
W	N	The other day.
W	P	The other night.

OTHER.—Continued.

J	W	Q	The other week.
J	W	R	The other one.
J	W	S	I have no other.

OVERBOARD.

J	W	T	Did you lose anything overboard?
J	W	V	Shall I throw him (them) overboard?
K	B	C	I lost it overboard.
K	B	D	Lost a man (men) overboard.
K	B	F	Throw it (them) overboard.
K	B	G	I will throw it (them) overboard.
K	B	H	Overboard they go if you say so.

P

PAPERS.

K	B	J	Have you any late papers?
K	B	L	What late papers have you?
K	B	M	Will you send the papers?
K	B	N	We have no late papers.
K	B	P	We have papers to the ———.

Yachting Code of Signals.

PAPERS.—Continued.

K B Q	We have the Herald.
K B R	" " " Times.
K B S	" " " Tribune.
K B T	" " " World.
K B V	" " " Sun.
K B W	" " " Journal of Commerce.
K C B	" " " Express.
K C D	" " " Evening Post.
K C F	We have late Boston papers.
K C G	Have sent you some late papers.
K C H	I will send for the papers.
K C J	The papers have been sent.
K C L	Bring some late papers with you.

PLEASE.

K C M	How can I please you?
K C N	If you (they) please.
K C P	If I please.
K C Q	Please yourself.
K C R	Please do so.
K C S	I do not please.

PLEASURE.

K C T	With much pleasure.
K C V	It will give me much pleasure.

POSTPONE—D.

K	C	W	Can it be postponed?
K	D	B	Has it been postponed?
K	D	C	Postpone it if possible.
K	D	F	It will be postponed.
K	D	G	It cannot be postponed.
K	D	H	It has been postponed.
K	D	J	It has not been postponed.
K	D	L	It will not be postponed.
K	D	M	Departure postponed until ———.

PREPARE—D.

K	D	N	Are you prepared?
K	D	P	We are prepared.
K	D	Q	We are not prepared.
K	D	R	Prepare as soon as possible.
K	D	S	When you are prepared.
K	D	T	Will be prepared in time.
K	D	V	Will not be prepared in time.

PREPARATION.

K	D	W	Are your preparations completed?
K	F	B	Our preparations are almost completed.
K	F	C	Make every preparation.
K	F	D	Every preparation is being made.
K	F	G	Preparations are completed.

Yachting Code of Signals.

PREPARING.

K	F	H	Are you preparing?
K	F	J	We are preparing.
K	F	L	We are not preparing.

PREVENT—ED.

K	F	M	Shall I prevent it?
K	F	N	Can it be prevented?
K	F	P	Why were you prevented?
K	F	Q	Prevent it if possible.
K	F	R	Do not prevent it.
K	F	S	It must be prevented.
K	F	T	It shall be prevented.
K	F	V	It has been prevented.
K	F	W	I was prevented.
K	G	B	I was unavoidably prevented.

Q

QUESTION.

K	G	C	Shall I ask the question?
K	G	D	What is the question?
K	G	F	Have you asked the question?

QUESTION.—Continued.

K	G	H	Does the question require an answer?
K	G	J	Ask the question.
K	G	L	The question must be asked.
K	G	M	The question requires an answer.
K	G	N	The question does not require an answer.

QUICKLY.

K	G	P	Was it done quickly?
K	G	Q	Will it be done quickly?
K	G	R	It must be done quickly.
K	G	S	It was done quickly.

R

RACE.

K	G	T	When will the race come off?
K	G	V	Did the race come off?
K	G	W	Who won the race?
K	H	B	Was it a race for a challenge cup?
K	H	C	When does the ocean race take place?
K	H	D	Who won the ocean race?
K	H	F	Was the race a fair one?

Yachting Code of Signals.

RACE.—Continued.

K	H	G	Was the race fairly won?
K	H	J	Will it be a fair race?
K	H	L	Do you enter for the race?
K	H	M	How many yachts have entered?
K	H	N	Can I enter for the race?
K	H	P	The race will come off on ———.
K	H	Q	The race came off on ———.
K	H	R	It was a race for a challenge cup.
K	H	S	It was not a race for a challenge cup.
K	H	T	The ocean race will take place on ———.
K	H	V	The ocean race was won by ———.
K	H	W	The race was won by ———.
K	J	B	The ocean race will not come off.
K	J	C	The race will not come off.
K	J	D	If the race comes off.
K	J	F	It will be an important race.
K	J	G	It will not be an important race.
K	J	H	Race postponed until ———.
K	J	L	The race was a fair one.
K	J	M	The race was not fair.
K	J	N	The race was fairly won.
K	J	P	The race was not fairly won.
K	J	Q	I think it will be a fair race.
K	J	R	I do not think it will be a fair race.
K	J	S	I will enter for the race.
K	J	T	I will not enter for the race.

RACE.—Continued.

K	J	V	Do not know whether I shall enter for the race.
K	J	W	The yachts that have entered number ———.
K	L	B	You can enter for the race.
K	L	C	You cannot enter for the race.
K	L	D	It is too late to enter.
K	L	F	There is time enough for you to enter.
K	L	G	It is a pity that he lost the race.
K	L	H	In the late race.

RAIN—ING—Y.

K	L	J	Shall we have rain?
K	L	M	Will the rain prevent?
K	L	N	Have you had much rain?
K	L	P	Was it raining?
K	L	Q	It will rain.
K	L	R	It will not rain.
K	L	S	It will not rain much.
K	L	T	The rain prevented.
K	L	V	The rain will prevent.
K	L	W	In the rain.
K	M	B	Not in the rain.
K	M	C	We have had very rainy weather.
K	M	D	The weather has not been rainy.
K	M	F	It was raining.
K	M	G	It was not raining.
K	M	H	On account of the rain.

RATHER.

K	M	J	Would you rather not?
K	M	L	Would you rather?
K	M	N	I would rather.
K	M	P	I would rather not.

REASON—S.

K	M	Q	What is (was) the reason?
K	M	R	Why not listen to reason?
K	M	S	There is (was) no reason.
K	M	T	I have (had) a good reason.
K	M	V	I have (had) no particular reason.
K	M	W	The reason is not sufficient.
K	N	B	The reason is satisfactory.
K	N	C	Give me a better reason.
K	N	D	I can give no other reason.
K	N	F	You do not reason correctly.
K	N	G	You reason very correctly.
K	N	H	I think my (our) reasons good and sufficient.
K	N	J	I must have better reasons next time.
K	N	L	Wait until you hear our reasons.
K	N	M	I will wait to hear your reasons.
K	N	P	You reason very badly.

REFLECT—ED—ION.

K	N	Q	Will you reflect?
K	N	R	Have you given the matter mature reflection?
K	N	S	Will it reflect upon me?
K	N	T	I will reflect upon it.
K	N	V	You must reflect upon it.
K	N	W	Reflect well upon it before you decide.
K	P	B	I have reflected well upon it.
K	P	C	Upon mature reflection.
K	P	D	He gave it mature reflection.
K	P	F	It will reflect upon me.
K	P	G	It will not reflect upon me.
K	P	H	It will reflect upon you.
K	P	J	It will not reflect upon you.

REST—ED—ING.

K	P	L	Are you taking a rest?
K	P	M	Are you rested yet?
K	P	N	We are taking a rest.
K	P	Q	You will have to rest.
K	P	R	After we are rested.
K	P	S	When you are rested.
K	P	T	We are resting after the fatigue.

RETURN—ED—ING.

K	P	V	When will you return?
K	P	W	Do you intend to return?
K	Q	B	Has he returned?
K	Q	C	Is he returning?
K	Q	D	I shall return soon.
K	Q	F	I shall not return until late.
K	Q	G	Return as soon as possible.
K	Q	H	I will return as soon as possible.
K	Q	J	Report on your return.
K	Q	L	I shall return at night.
K	Q	M	I shall return.
K	Q	N	I shall not return.
K	Q	P	I hope you will return.
K	Q	R	He has returned.
K	Q	S	He has not returned.
K	Q	T	When he has returned.
K	Q	V	He is returning.
K	Q	W	He is not returning.
K	R	B	Return to the port from which we started.

RIGHT—ED.

K	R	C	Am I right?
K	R	D	Who is right?
K	R	F	Which is right?
K	R	G	What is right?

RIGHT—ED.—Continued.

K	R	H	Will you see me righted?
K	R	J	You are right.
K	R	L	They are right.
K	R	M	You are not right.
K	R	N	They are not right.
K	R	P	If he is in the right.
K	R	Q	I think he is in the right.
K	R	S	You will be righted.
K	R	T	You must see him righted.

ROCKS.

K	R	V	Are the rocks dangerous?
K	R	W	The rocks are dangerous, look out for them.
K	S	B	The rock are not dangerous.
K	S	C	The rocks are awash at low water.
K	S	D	The rocks are covered at high water.
K	S	F	Keep a good lookout for the rocks.
K	S	G	I am keeping a bright lookout for the rocks.
K	S	H	If you (when you) sight the rocks report to me.
K	S	J	The rocks are in sight.

RUDDER.

K	S	L	Is anything the matter with your rudder?
K	S	M	My rudder is broken.

Yachting Code of Signals.

RUN—NING.

K	S	N	Shall I run for it?
K	S	P	Did you make a good run?
K	S	Q	Where are you running?
K	S	R	You will have to run for it.
K	S	T	Run in and make the land.
K	S	V	I made a good run.
K	S	W	I made a bad run.
K	T	B	The run was very good.
K	T	C	The run was bad.
K	T	D	I am running to make the land.
K	T	F	I am running to make a harbor.
K	T	G	I am running to join the squadron.

S

SAFE—LY.

K	T	H	Are you all safe?
K	T	J	Have you returned safely?
K	T	L	Is it (will it be) a safe operation?
K	T	M	We are all safe.
K	T	N	We have returned safely.
K	T	P	I am glad to hear of your safe return.
K	T	Q	I do not think it will be safe.

Yachting Code of Signals.

SAFE—LY.—Continued.

K	T	R	You can do it safely.
K	T	S	I wish you a safe return.

SAIL.

K	T	V	Shall I make sail?
K	T	W	When will you sail?
K	V	B	When do you wish us to sail?
K	V	C	When shall we sail?
K	V	D	When will you be ready to sail?
K	V	F	Do you think we sail fast?
K	V	G	What do you make of the sail (sails) in sight?
K	V	H	Shall I (we) shorten sail?
K	V	J	I shall sail at ——— (on ———).
K	V	L	I shall not sail for some time.
K	V	M	You will sail as soon as the signal is made.
K	V	N	Make sail.
K	V	P	Make sail for ———.
K	V	Q	When ready to sail report by signal.
K	V	R	I am ready to sail.
K	V	S	The yacht sails very well.
K	V	T	You appear to sail very fast.
K	V	W	The sail (sails) in sight is (are) a yacht (yachts).
K	W	B	The sail (sails) in sight is (are) a merchant vessel (merchant vessels).
K	W	C	The sail (sails) in sight is (are) a man-(men)of-war.
K	W	D	A sail (sails) in sight ahead.

SAIL.—Continued.

K	W	F	A sail (sails) in sight **astern**.
K	W	G	A sail (sails) in sight to **starboard**.
K	W	H	A sail (sails) in sight to **port**.
K	W	J	Do not make more sail.
K	W	L	Shorten sail.
K	W	M	Make more sail.
K	W	N	Do not make more sail.

SAILS.

K	W	P	My sails need repairing.
K	W	Q	My ——— is (are) disabled.
K	W	R	Must put back to repair sails.
K	W	S	Can you lend me a ———?
K	W	T	Must remain to repair sails.
K	W	V	Prepare to dry sails.
L	B	C	Dry sails.
L	B	D	Our sails need drying, shall we loose them?
L	B	F	Do your sails need drying?
L	B	G	Jibtopsail.
L	B	H	Flying jib.
L	B	J	Jib.
L	B	K	Forestaysail.
L	B	M	Foresail.
L	B	N	Mainsail.
L	B	P	Maintopsail.

SAILS.—Continued.

L	B	Q	Foretopsail.
L	B	R	Fore-club-topsail.
L	B	S	Main-club-topsail.
L	B	T	Maintopmast-staysail.
L	B	V	Balloon jib.

SALUTE.

L	B	W	Will a salute be fired?
L	C	B	When will a salute be fired?
L	C	D	At what hour will the salute be fired?
L	C	F	Of how many guns will the salute consist?
L	C	G	A salute will be fired.
L	C	H	No salute will be fired.
L	C	J	Prepare to fire a salute.
L	C	K	Fire a salute of ——— guns on signal.
L	C	M	Commence firing.

SAME.

L	C	N	Is it the same?
L	C	P	It is the same.
L	C	Q	It is not the same.
L	C	R	At the same time.

Yachting Code of Signals.

SAVE—D.

L	C	S	Will it save time?
L	C	T	Can anything (any time) be saved?
L	C	V	It will save time.
L	C	W	It will not save time.
L	D	B	Much can be saved.
L	D	C	Nothing can be saved.
L	D	F	If you wish to save time.
L	D	G	Save all the time you can.
L	D	H	He can be saved if you work quickly.
L	D	J	He has been saved.
L	D	K	He was not saved.

SAY.

L	D	M	Have you anything to say?
L	D	N	We have nothing to say.
L	D	P	I have much to say to you.
L	D	Q	If you have anything to say.

SEA.

L	D	R	Have you had a rough sea?
L	D	S	How is the sea outside?
L	D	T	The sea has been very rough.
L	D	V	The sea is (was) quite smooth.
L	D	W	There is a rough sea outside.

SEA.—Continued.

L	F	B	A heavy head sea.
L	F	C	A heavy beam sea.

SEE—N.—(See "Come.")

L	F	D	Who (what) did you see?
L	F	G	Have you seen him?
L	F	H	When do you expect to see him?
L	F	J	When will you come to see me?
L	F	K	I did not see anything (anybody).
L	F	M	I have seen nothing.
L	F	N	I do not expect to see him.
L	F	P	I expect to see him on ———.
L	F	Q	Come and see me.
L	F	R	I am coming to see you.
L	F	S	I cannot come to see you.

SHALL.

L	F	T	Shall I?
L	F	V	I shall do so.
L	F	W	You shall not do so.
L	G	B	You shall.

SICK—NESS.

L	G	C	Are any of you sick?
L	G	D	Have you sickness aboard?

SICK—NESS.—Continued.

L	G	F	None of us are sick.
L	G	H	We have sickness aboard.
L	G	J	We have no sickness aboard.
L	G	K	I hope that none of you are sick.

SIGHT.—(See "Land.")

L	G	M	Was it a fine sight?
L	G	N	Did you sight the land?
L	G	P	Is the land in sight?
L	G	Q	What is in sight?
L	G	R	Is anything in sight?
L	G	S	It was a beautiful sight.
L	G	T	It was not much of a sight.
L	G	V	We sighted the land.
L	G	W	We did not sight the land.
L	H	B	The land is in sight.
L	H	C	The land is not in sight.
L	H	D	When you sight the land, report by signal.

SIGNAL—S—ED.

L	H	F	Shall I make signal to the ———?
L	H	G	What signal does she show?
L	H	J	Does she make any signal?
L	H	K	Can you make out the signal?

Yachting Code of Signals.

SIGNAL—S—ED.—Continued.

L	H	M	Shall I repeat the signal?
L	H	N	Have you signaled anything?
L	H	P	Is she within signal distance?
L	H	Q	She is showing Marryatt's signals.
L	H	R	She is showing Rogers' signals.
L	H	S	I cannot read your signals.
L	H	T	Your signal does not blow out clear.
L	H	V	Repeat my signals.
L	H	W	Repeat my signals to the lee division.
L	J	B	Repeat my signals to the weather division.
L	J	C	Pay more attention to the signals.
L	J	D	The signals are easily distinguished.
L	J	F	Prepare to exercise signals with the Yachting Code.
L	J	G	Answer the signals promptly.
L	J	H	Report by signal.
L	J	K	We have signaled nothing.
L	J	M	We signaled the squadron.
L	J	N	Make signal to the ———.
L	J	P	She is within signal distance.
L	J	Q	She is not within signal distance.
L	J	R	Make signal to her.
L	J	S	Set your signals.

SINCE.

L	J	T	How long since?
L	J	V	Not long since.

SINCE.—Continued.

L	J	W	Some time since.
L	K	B	Since I last saw you.

SLIP—PED.

L	K	C	Shall we slip?
L	K	D	Has she (have they) slipped?
L	K	F	Slip your chain.
L	K	G	Slip as soon as possible.
L	K	H	She has (they have) not slipped.
L	K	J	She has (they have) slipped.
L	K	M	I will slip.
L	K	N	You will have to slip.
L	K	P	I shall have to slip.

SO.

L	K	Q	Is it so?
L	K	R	Did you say so?
L	K	S	It is so.
L	K	T	It is not so.
L	K	V	I said so.
L	K	W	I did not say so.

SOUNDINGS.

L	M	B	Have you had soundings?
L	M	C	What soundings have you?

SOUNDINGS.—Continued.

L	M	D	What soundings have you had?
L	M	F	We are on soundings.
L	M	G	We are off soundings.

SOUTH—ERLY—WARD.

L	M	H	Have you had southerly winds?
L	M	J	Is anything in sight to the southward?
L	M	K	Shall I stand to the southward?
L	M	N	Shall I pass to the southward?
L	M	P	Off the south side of ———.
L	M	Q	Off the south end of ———.
L	M	R	Southerly.
L	M	S	We have had southerly winds.
L	M	T	We have had a strong southerly current.
L	M	V	I think you will have southerly winds.
L	M	W	On a southerly course.
L	N	B	In a southerly direction.

STEAM—ER—S.

L	N	C	Can you send me a steam tug?
L	N	D	Do you want a steam tug?
L	N	F	Is she a steamer?
L	N	G	Have you seen any steamers?
L	N	H	I will send you a (the) steam tug.

Yachting Code of Signals.

STEAM—ER—S.—Continued.

L	N	J	I am in want of a steam tug.
L	N	K	She is a steamer.
L	N	M	Steam yacht.
L	N	P	We have seen several steamers.
L	N	Q	We have seen no steamers.
L	N	R	Look out for a steamer (steamers).

SUPPER.

L	N	S	Will you have a supper to-night?
L	N	T	Will you join us at supper.
L	N	V	At what hour do you take supper?
L	N	W	We shall have a supper to-night.
L	P	B	Will join you at supper.
L	P	C	We take supper at ———.
L	P	D	We cannot stand late suppers.
L	P	F	Supper is ready.

SUPPLY—IES.

L	P	G	Can we get supplies at ———?
L	P	H	You can get supplies at ———.
L	P	J	I am waiting for supplies.
L	P	K	We are short of supplies.
L	P	M	We can supply you.
L	P	N	We have no supplies to spare.
L	P	Q	Have sent ashore for supplies.

SURGEON.

L	P	R	Have you a surgeon on board?
L	P	S	Is there a surgeon in the squadron?
L	P	T	Do you want a surgeon?
L	P	V	When will the surgeon return?
L	P	W	There is a surgeon on the ———.
L	Q	B	We want a surgeon immediately.
L	Q	C	There is a surgeon ashore.
L	Q	D	The surgeon is absent.

T

TACK.—(See "Wrong.")

L	Q	F	Shall we tack ship?
L	Q	G	Tack ship.
L	Q	H	Do not tack ship.
L	Q	J	I will tack ship.

TALK.

L	Q	K	Did it occasion much talk?
L	Q	M	Was it talked about?
L	Q	N	It occasioned a great deal of talk.

TALK.—Continued.

L	Q	P	It did not occasion much talk.
L	Q	R	We will talk the matter over.
L	Q	S	It will be talked about.
L	Q	T	It will have to be talked about.

TELL.

L	Q	V	Will you tell him?
L	Q	W	I will not tell him.
L	R	B	I shall have to tell him.
L	R	C	I will tell you when we next meet.

THINK—ING.

L	R	D	Do you think so?
L	R	F	What do you think of it?
L	R	G	Do you think I am right?
L	R	H	When will you think of it?
L	R	J	I think so.
L	R	K	I do not think so.
L	R	M	I think nothing of it.
L	R	N	I have been thinking of it.
L	R	P	I think you are right.
L	R	Q	I think you are wrong.
L	R	S	I will think of it shortly.
L	R	T	You will think differently.

TIDE.

L	R	V	How is the tide?
L	R	W	It is high tide.
L	S	B	It is low tide.
L	S	C	It is half tide.
L	S	D	It is a rising tide.
L	S	F	It is a falling tide.
L	S	G	It is slack tide.
L	S	H	It is a strong tide.
L	S	J	It is a very strong tide.

TIME.

L	S	K	What time is it?
L	S	M	Is it about time?
L	S	N	What time do you propose?
L	S	P	Will you be in time?
L	S	Q	It is about time.
L	S	R	I shall be in time.
L	S	T	The time has passed.
L	S	V	It is time he was here.
L	S	W	You will time the yachts.
L	T	B	You must be in time.
L	T	C	Immediately (now, at once).

U

UNABLE.

L	T	D	Are you unable to attend (be there)?
L	T	F	Why were you unable?
L	T	G	I am unable to attend (be there).
L	T	H	I am sorry you are unable.
L	T	J	If you are unable, let me know.

UNDERWEIGH.

L	T	K	Is she (are they) underweigh?
L	T	M	When will you get underweigh?
L	T	N	Shall I get underweigh?
L	T	P	How shall I steer when underweigh?
L	T	Q	Was she (were they) underweigh?
L	T	R	Was the squadron underweigh?
L	T	S	We shall get underweigh on signal.
L	T	V	We are not ready to get underweigh.
L	T	W	We are ready to get underweigh.
L	V	B	The squadron will get underweigh at ——— o'clock.
L	V	C	She is (they are) underweigh.
L	V	D	She is (they are) not underweigh.
L	V	F	She is (they are) going to get underweigh.

UNDERWEIGH.—Continued.

L	V	G	When underweigh, follow my movements.
L	V	H	When underweigh you will steer ———.
L	V	J	When underweigh we shall steer ———.
L	V	K	Get underweigh as soon as possible.
L	V	M	Be ready to get underweigh on signal.
L	V	N	The squadron was underweigh.

UNWILLING.

L	V	P	Are you unwilling?
L	V	Q	We are unwilling.
L	V	R	Do not be unwilling.

USE—FUL.

L	V	S	Is it of use?
L	V	T	Will I be useful?
L	V	W	It is of use.
L	W	B	It is of no use.
L	W	C	I will make myself useful.
L	W	D	You will be very useful.

WANT.

L	W	F	What do you want?
L	W	G	Do you want me?
L	W	H	We want nothing.
L	W	J	I want you.
L	W	K	If you want anything let me know.

WAS.

L	W	M	Was he?
L	W	N	I was.
L	W	P	I was not.

WATER.

L	W	Q	Is the water fresh?
L	W	R	How much water have you?
L	W	S	What water is there where you are anchored?
L	W	T	Is there water enough?
L	W	V	I am in want of water.
M	B	C	I can spare you some water.
M	B	D	There is water enough.
M	B	F	There is not water enough.

WAY.

M	B	G	Will you show the way in?
M	B	H	I will show you the way.
M	B	J	I cannot show you the way.

WEAR.

M	B	K	Shall I wear ship?
M	B	L	Wear ship.
M	B	N	Do not wear ship.
M	B	P	I am going to wear ship.

WEATHER.

M	B	Q	Do you think we shall have fine weather?
M	B	R	Do you think we shall have bad weather?
M	B	S	Do you think we shall have foggy weather?
M	B	T	Do you think we shall have heavy weather?
M	B	V	Do you think we shall have a gale?
M	B	W	Do you think we shall have a storm?
M	C	B	Do you think we shall have a squall?
M	C	D	Do you think we shall have a calm?
M	C	F	Have you had fine weather?
M	C	G	Have you had bad weather?
M	C	H	Have you had foggy weather?
M	C	J	Have you had heavy weather?
M	C	K	Have you had a gale?

Yachting Code of Signals.

WEATHER.—Continued.

M	C	L	Have you had a storm?
M	C	N	Have you had a squall?
M	C	P	Have you had a calm?
M	C	Q	We expect fine weather.
M	C	R	We expect bad weather.
M	C	S	We expect foggy weather.
M	C	T	We expect heavy weather.
M	C	V	We expect a gale.
M	C	W	We expect a storm.
M	D	B	We expect a squall.
M	D	C	We expect a calm.
M	D	F	We have had fine weather.
M	D	G	We have had heavy weather.
M	D	H	We have had bad weather.
M	D	J	We have had foggy weather.
M	D	K	We have had a gale.
M	D	L	We have had a storm.
M	D	N	We have had a calm.
M	D	P	We have had a squall.
M	D	Q	There is no indication of fine weather.
M	D	R	There is no indication of bad weather.
M	D	S	There is no indication of heavy weather.
M	D	T	There is no indication of foggy weather.
M	D	V	There is no indication of a gale.
M	D	W	There is no indication of a storm.
M	F	B	There is no indication of a squall.

WEATHER.—Continued.

M	F	C	There is no indication of a calm.
M	F	D	Hope you will have fine weather.
M	F	G	Fear you will have bad weather.
M	F	H	Fear you will have heavy weather.
M	F	J	Fear you will have foggy weather.
M	F	K	Fear you will have a gale.
M	F	L	Fear you will have a storm.
M	F	N	Fear you will have a squall.
M	F	P	Fear you will have a calm.
M	F	Q	On account of the storm.
M	F	R	On account of the calm.
M	F	S	On account of the fog.
M	F	T	If the weather is fine.
M	F	V	If the weather is bad.
M	F	W	There are indications of a heavy gale, make a harbor at ———.
M	G	B	The gale is too heavy to run, we must heave to.
M	G	C	We can ride out the gale in safety.

WEST—ERLY—WARD.

M	G	D	Off the west side of ———.
M	G	F	Off the west end of ———.
M	G	H	To the westward of ———.
M	G	J	We have had westerly winds.
M	G	K	We have had a westerly current.
M	G	L	You will probably have westerly winds.

Yachting Code of Signals.

WEST—ERLY—WARD.—Continued.

M	G	N	Have you had westerly winds?
M	G	P	Is anything in sight to the westward?
M	G	Q	Shall I stand to the westward?
M	G	R	Shall I pass to the westward?
M	G	S	Westerly.
M	G	T	On a westerly course.
M	G	V	In a westerly direction.

WHERE.

M	G	W	Where is the ———?

WHY.

M	H	B	Why?
M	H	C	Why not?

WILL—ING.

M	H	D	Will you?
M	H	F	Are you willing?
M	H	G	I will.
M	H	J	I will not.
M	H	K	I am willing.
M	H	L	I am not willing.
M	H	N	If you are willing.

WIND.

M	H	P	How have you had the wind?
M	H	Q	Has the wind been light?
M	H	R	Will there be a change of wind?
M	H	S	Have you had variable winds?
M	H	T	We have had the wind from ———.
M	H	V	The wind has been light.
M	H	W	The wind has been strong.
M	J	B	The wind has been variable.
M	J	C	There will be a change of wind.
M	J	D	By the wind.
M	J	F	Before the wind.
M	J	G	Off on the wind.

WRONG.—(See "Think.")

M	J	H	Are we wrong?
M	J	K	Was it wrong?
M	J	L	You are wrong.
M	J	N	It was wrong.
M	J	P	You are on the wrong tack.

YACHT.

M	J	Q	What yacht is that?
M	J	R	What yachts have you seen?
M	J	S	Does the yacht work well?

YACHT.—Continued.

M	J	T	The yacht is working badly.
M	J	V	The yacht is aground.
M	J	W	The yacht is afloat.

YEAR.

M	K	B	What year?
M	K	C	This year.
M	K	D	Last year.
M	K	F	Next year.
M	K	G	Not for years.

YOU—R—S.

M	K	H	Is it you?
M	K	J	How are you?
M	K	L	When will you?
M	K	N	Will you?
M	K	P	Shall I use your name?
M	K	Q	Shall I use your boat?
M	K	R	Good for you.
M	K	S	It is not yours.
M	K	T	In your name.
M	K	V	You will use my name.

Yachting Code of Signals.

Yachting Code of Signals.

YACHT NUMBERS.

SCHOONERS.

N	B	C	
N	B	D	Alarm.
N	B	F	Ariel.
N	B	G	Atalanta.
N	B	H	
N	B	J	
N	B	K	
N	B	L	
N	B	M	
N	B	P	
N	B	Q	Clio.
N	B	R	Columbia.
N	B	S	Comet.
N	B	T	Cornelia.
N	B	V	
N	B	W	
N	C	B	
N	C	D	Dauntless.
N	C	F	Dreadnaught.
N	C	G	
N	C	H	
N	C	J	
N	C	K	Edith.
N	C	L	Enchantress.
N	C	M	Eva.

Yachting Code of Signals.

YACHT NUMBERS.—Continued.

SCHOONERS.—Continued.

N	C	P	
N	C	Q	
N	C	R	
N	C	S	Faustine.
N	C	T	Fleetwing.
N	C	V	Fleur de Lis.
N	C	W	Foam.
N	D	B	
N	D	C	
N	D	F	
N	D	G	Gypsie.
N	D	H	
N	D	J	
N	D	K	
N	D	L	
N	D	M	
N	D	P	
N	D	Q	Ibis.
N	D	R	Idler.
N	D	S	
N	D	T	
N	D	V	
N	D	W	Josephine.
N	F	B	
N	F	C	

YACHT NUMBERS.—Continued.

SCHOONERS.—Continued.

N	F	D	
N	F	G	
N	F	H	
N	F	J	
N	F	K	Madeleine.
N	F	L	Magic.
N	F	M	
N	F	P	
N	F	Q	
N	F	R	
N	F	S	
N	F	T	
N	F	V	Palmer.
N	F	W	Peerless.
N	G	B	Phantom.
N	G	C	
N	G	D	
N	G	F	
N	G	H	Rambler.
N	G	J	Rebecca.
N	G	K	Resolute.
N	G	L	Restless.
N	G	M	
N	G	P	
N	G	Q	

YACHT NUMBERS.—Continued.

SCHOONERS.—Continued.

N	G	R	Sappho.
N	G	S	Seadrift.
N	G	T	Sunshine.
N	G	V	Swan.
N	G	W	
N	H	B	
N	H	C	
N	H	D	Tarolinta.
N	H	F	Tidal Wave.
N	H	G	
N	H	J	
N	H	K	
N	H	L	Vesta.
N	H	M	Viking.
N	H	P	
N	H	Q	
N	H	R	
N	H	S	Wanderer.
N	H	T	
N	H	V	
N	H	W	

YACHT NUMBERS.—Continued.

SLOOPS.

N	J	B	Aida.
N	J	C	Alert.
N	J	D	Alice.
N	J	F	Ariadne.
N	J	G	
N	J	H	
N	J	K	
N	J	L	Breeze.
N	J	M	
N	J	P	
N	J	Q	
N	J	R	Christine.
N	J	S	
N	J	T	
N	J	V	
N	J	W	Dudley.
N	K	B	
N	K	C	
N	K	D	
N	K	F	Elaine.
N	K	G	
N	K	H	
N	K	J	
N	K	L	Fanny.
N	K	M	

Yachting Code of Signals.

YACHT NUMBERS.—Continued.

SLOOPS.—Continued.

N	K	P	
N	K	Q	
N	K	R	Genia.
N	K	S	Gracie.
N	K	T	
N	K	V	
N	K	W	
N	L	B	
N	L	C	
N	L	D	
N	L	F	Irene.
N	L	G	
N	L	H	
N	L	J	
N	L	K	Josie.
N	L	M	
N	L	P	
N	L	Q	
N	L	R	Kate.
N	L	S	
N	L	T	
N	L	V	
N	L	W	
N	M	B	
N	M	C	Qui Vive.

YACHT NUMBERS.—Continued.

SLOOPS.—Continued.

N M D
N M F
N M G   Sallie E. Day.
N M H
N M J
N M K
N M L
N M P
N M Q
N M R   Vindex.
N M S   Vision.
N M T   Vixen.
N M V
N M W
N P B   Wayward.
N P C
N P D
N P F
N P G

STEAMERS.

N P H
N P J
N P K
N P L
N P M   Day Dream.

Yachting Code of Signals.

YACHT NUMBERS.—Continued.

STEAMERS.—Continued.

N	P	Q	
N	P	R	
N	P	S	Emily.
N	P	T	
N	P	V	
N	P	W	Fearless.
N	Q	B	
N	Q	C	
N	Q	D	
N	Q	F	Ideal.
N	Q	G	
N	Q	H	
N	Q	J	Julia.
N	Q	K	
N	Q	L	Lady of the Lake.
N	Q	M	Lurline.
N	Q	P	
N	Q	R	Mystic.
N	Q	S	
N	Q	T	
N	Q	V	Wave.
N	Q	W	Wyvern.

Yachting Code of Signals.

NAMES OF PLACES.

P	B	C	Abaco, Bahamas.
P	B	D	Absecom, N. J.
P	B	F	Amboy, N. J.
P	B	G	Annapolis, Md.
P	B	H	Baltimore, Md.
P	B	J	Barbadoes, W. I.
P	B	K	Barnegat, N. J.
P	B	L	Beaufort, S. C.
P	B	M	Beaver Tail.
P	B	N	Bermuda.
P	B	Q	Black Point.
P	B	R	Black Rock.
P	B	S	Block Island.
P	B	T	Bodies Island, N. C.
P	B	V	Boston.
P	B	W	Branford.
P	C	B	Bridgeport.
P	C	D	Caldwells, N. R.
P	C	F	Cape Ann, Mass.
P	C	G	Cape Carnaveral, Florida.
P	C	H	Cape Charles, Va.
P	C	J	Cape Cod, Mass.
P	C	K	Cape Fear, N. C.
P	C	L	Cape Florida.
P	C	M	Cape Hatteras, N. C.
P	C	N	Cape Henry, Va.

Yachting Code of Signals.

NAMES OF PLACES.—Continued.

P	C	Q	Cape Henlopen, Del.
P	C	R	Cape Lookout, N. C.
P	C	S	Cape May, N. J.
P	C	T	Cape Maysi, east end of Cuba.
P	C	V	Cape San Antonio, west end of Cuba.
P	C	W	Captain's Islands.
P	D	B	Cedar Keys, Fla.
P	D	C	Charleston, S. C.
P	D	F	Charles Island.
P	D	G	Clark's Point, Buzzard Bay.
P	D	H	Cold Spring, L. I.
P	D	J	Cold Spring, N. R.
P	D	K	Cornwall.
P	D	L	Cuba.
P	D	M	Cutty Hunk Harbor.
P	D	N	Delaware Breakwater.
P	D	Q	Edgartown.
P	D	R	Falkner's Island.
P	D	S	Fernandina, Fla.
P	D	T	Fire Island.
P	D	V	Fishkill, N. R.
P	D	W	Five fathom bank Light-ship, off Delaware Bay.
P	F	B	Fort Bend Bay.
P	F	C	Fort Lee, N. R.
P	F	D	Fryingpan Shoals, N. C.
P	F	G	Galveston.

NAMES OF PLACES.—Continued.

P	F	H	Gardner's Island.
P	F	J	Garrison's, N. R.
P	F	K	George's Shoals.
P	F	L	Glen Cove.
P	F	M	Gravesend Bay.
P	F	N	Greenpoint.
P	F	Q	Greenport.
P	F	R	Hallet's Cove.
P	F	S	Hampton Roads.
P	F	T	Harlem River.
P	F	V	Hart Island.
P	F	W	Hastings, N. R.
P	G	B	Havana.
P	G	C	Hoboken.
P	G	D	Hog Island.
P	G	F	Hole in the Wall, east end of Abaco, Bahamas.
P	G	H	Holme's Hole.
P	G	J	Horse Shoe, Sandy Hook.
P	G	K	Horton's Point.
P	G	L	Hyannis.
P	G	M	Jamaica, W. I.
P	G	N	Keyport.
P	G	Q	Keywest, Fla.
P	G	R	Kingston, Jamaica.
P	G	S	Kinsale, Old Head of.
P	G	T	Lloyds Harbor.

Yachting Code of Signals.

NAMES OF PLACES.—Continued.

P	G	V	Long Branch.
P	G	W	Marblehead, Mass.
P	H	B	Martha's Vineyard.
P	H	C	Martinique, W. I.
P	H	D	Matanzas.
P	H	F	Mobile.
P	H	G	Montauk.
P	H	J	Nahant.
P	H	K	Nantucket.
P	H	L	Nantucket Shoals.
P	H	M	Napeag Harbor.
P	H	N	Nassau, N. P., Bahamas.
P	H	Q	New Bedford.
P	H	R	Newbern, N. C.
P	H	S	New Brighton, S. I.
P	H	T	Newburgh, N. R.
P	H	V	Newburyport, Mass.
P	H	W	New Haven.
P	J	B	New London.
P	J	C	Newport.
P	J	D	New Rochelle.
P	J	F	Norfolk, Va.
P	J	G	Nyack, N. R.
P	J	H	Oak Bluffs.
P	J	K	Orient.
P	J	L	Oyster Bay.

Yachting Code of Signals.

NAMES OF PLACES.—Continued.

P	J	M	Oyster Pond Point.
P	J	N	Pensacola, Fla.
P	J	Q	Philadelphia.
P	J	R	Point Judith.
P	J	S	Ponce, Island of Porto Rico.
P	J	T	Port Jefferson.
P	J	V	Portland, Me.
P	J	W	Porto Rico, W. I.
P	K	B	Port Royal, Jamaica.
P	K	C	Portsmouth, N. H.
P	K	D	Prince's Bay.
P	K	F	Providence, R. I.
P	K	G	Provincetown.
P	K	H	Quarantine.
P	K	J	Quick's Hole.
P	K	L	Ram Island.
P	K	M	Riker's Island.
P	K	N	Rockaway Inlet.
P	K	Q	Rye Point.
P	K	R	Sachem's Head.
P	K	S	Sag Harbor.
P	K	T	Salem, Mass.
P	K	V	Samana Bay, San Domingo.
P	K	W	Sand Key, Fla.
P	L	B	San Domingo.
P	L	C	Savannah, Ga.

NAMES OF PLACES.—Continued.

P	L	D	Saybrook.
P	L	F	Seconnet.
P	L	G	Sheffield's Island.
P	L	H	Sing Sing, N. R.
P	L	J	Smithtown Bay, L. I.
P	L	K	Santiago de Cuba.
P	L	M	Stonington.
P	L	N	Stony brook.
P	L	Q	Stratford.
P	L	R	St. Thomas, W. I.
P	L	S	Tampa Bay, Fla.
P	L	T	Tarpaulin Cove.
P	L	V	Thimble Island.
P	L	W	Throgg's Neck.
P	M	B	Tortugas, Fla.
P	M	C	Trinidad Port Spain, W. I.
P	M	D	Vineyard Haven.
P	M	F	Warsaw Sound, Ga.
P	M	G	West Harbor, Fisher's Island.
P	M	H	West Point, N. R.
P	M	J	Whitestone.
P	M	K	Wilmington, N. C.
P	M	L	Woods Hole.
P	M	N	Yonkers.

www.ingramcontent.com/pod-product-compliance
Lightning Source LLC
Chambersburg PA
CBHW030255170426
43202CB00009B/750